HSE

Health and Safety Executive

Dangerous substances and explosive atmospheres

Dangerous Substances and Explosive Atmospheres Regulations 2002

Approved Code of Practice and guidance

London: TSO

HSE

tso
a Williams Lea company

Published by TSO (The Stationery Office), part of Williams Lea,
and available from:

Online
www.tsoshop.co.uk

Mail, Telephone, Fax & E-mail
TSO
PO Box 29, Norwich, NR3 1GN
Telephone orders/General enquiries: 0333 202 5070
Fax orders: 0333 202 5080
E-mail: customer.services@tso.co.uk
Textphone 0333 202 5077

TSO@Blackwell and other Accredited Agents

First published 2003
Second edition 2013

ISBN 978 0 7176 6616 4

Printed in the United Kingdom for The Stationery Office.
J003661517 c1 04/20

Approved Code of Practice

This Code has been approved by the Health and Safety Executive, with the consent of the Secretary of State. It gives practical advice on how to comply with the law. If you follow the advice you will be doing enough to comply with the law in respect of those specific matters on which the Code gives advice. You may use alternative methods to those set out in the Code in order to comply with the law.

However, the Code has a special legal status. If you are prosecuted for breach of health and safety law, and it is proved that you did not follow the relevant provisions of the Code, you will need to show that you have complied with the law in some other way or a Court will find you at fault.

Guidance

This guidance is issued by the Health and Safety Executive. Following the guidance is not compulsory, unless specifically stated, and you are free to take other action. But if you do follow the guidance you will normally be doing enough to comply with the law. Health and safety inspectors seek to secure compliance with the law and may refer to this guidance.

Contents

Introduction

About this book

1 The Dangerous Substances and Explosive Atmospheres Regulations 2002 (DSEAR) Approved Code of Practice (ACOP) text and associated guidance provide practical advice on how you can comply with the requirements of the Regulations.

2 The ACOP is intended primarily for an informed and experienced audience such as professional health and safety staff and those who may give advice to smaller businesses rather than the small businesses themselves. The leaflet INDG370[1] provides a short guide to DSEAR and is aimed at small and medium-sized businesses. Information on DSEAR can also be accessed via the Fire and explosion pages of HSE's website at www.hse.gov.uk/fireandexplosion.

3 This publication is for places of work that manufacture, store, process or use dangerous substances as defined in paragraph 33. It contains an ACOP explaining what is required for compliance, together with guidance on the duties in DSEAR. It has been prepared by the Health and Safety Executive (HSE) after consultation with stakeholders in industry, trades unions, local authorities and fire authorities. While *Unloading petrol from road tankers*[2] continues as a separate ACOP, four DSEAR ACOP publications from 2003 have been merged into the original main ACOP L138 *Dangerous Substances and Explosive Atmospheres* namely:

(a) L134 *Design of plant, equipment and workplaces*;
(b) L135 *Storage of dangerous substances;*
(c) L136 *Control and mitigation measures*;
(d) L137 *Safe maintenance, repair and cleaning procedures*.

4 The ACOP text and guidance have been simplified, streamlined and any previous ambiguities removed. The changes, which are summarised below, have been widely consulted on.

5 Changes in this edition include:

(a) some guidance has been assigned ACOP status and vice versa. For example, previously no ACOP material existed for regulations 7 and 10. However, these changes were introduced to provide clarification on how to comply and do not introduce any new requirements;
(b) adjustments in light of European and other legislation that has or will be introduced after DSEAR came into force such as:
    (i) general fire safety legislation (see paragraph 20);
    (ii) classification for labelling and packaging (see Appendix 3);
(c) clarification on the scope of the Regulations in relation to incompatible substances being outside the scope of DSEAR but within the scope of the Health and Safety at Work etc Act (the HSW Act) (see paragraph 255);
(d) Schedules to the Regulations have been moved from the end of the document to sit near their respective regulation;
(e) the list of references and further reading has been updated.

About ACOPs

6 Approved Codes of Practice are approved by the HSE Board with the consent of the Secretary of State (see Appendix 1: Notice of Approval for details).

7 The ACOP describes preferred or recommended methods that can be used (or standards to be met) to comply with the Regulations and the duties imposed by the HSW Act. The accompanying guidance also provides advice on achieving compliance, or it may give information of a general nature, including explanation of the requirements of the law, more specific technical information or references to further sources of information.

8 The legal status of ACOP and guidance text is given on the copyright page.

Presentation of regulation, guidance and ACOP text

9 The ACOP text is set out in **bold** and the accompanying guidance in normal type, the text of the Regulations is in *italics*. Coloured borders also indicate each section clearly. Each regulation is preceded by a short summary of the main duties imposed by that regulation and aims to help the reader navigate the document. This summary text is for information only.

About DSEAR

10 The Dangerous Substances and Explosive Atmospheres Regulations 2002 (SI 2002/2776) set minimum requirements for the protection of workers from fire and explosion risks related to dangerous substances and potentially explosive atmospheres. The Regulations apply to employers and the self-employed at most workplaces in Great Britain where a dangerous substance is present or could be present.

11 DSEAR revoked or modified a large amount of old legislation relating to flammable substances and dusts including the Highly Flammable Liquids and Liquefied Petroleum Gases Regulations 1972 (SI 1972/917) and section 31 of the Factories Act 1961. Safety standards were maintained through a combination of the requirements of DSEAR and ACOPs reflecting practices in the preceding legislation.

Scope of 'employer'

12 The term 'employer' is used throughout the document and should be understood to include the self-employed as a consequence of regulation 4(2). Less frequently the less-specific term 'dutyholder' may also be used which includes both.

Information for use in risk assessment

13 A key part of compliance with DSEAR is the risk assessment and an important principle is to separate storage areas from process areas where the two occur on the same or shared premises. Employers will find generic information on the storage of dangerous substances in other HSE and non-HSE guidance. In contrast process activities are, by their nature, very variable, and the employer should consider these in much more detail and apply the principles of risk assessment themselves as there may be less guidance available.

14 Health effects from substances and preparations are not within the scope of DSEAR and are covered by legislation such as the Control of Substances Hazardous to Health Regulations 2002 (SI 2002/2677) (COSHH).

The legislative background

15 DSEAR 2002 was made under the Health and Safety at Work etc Act 1974 (the HSW Act). The Regulations apply to workplaces in Great Britain. The Regulations implement two European Directives:

(a) the safety aspects of the Chemical Agents Directive 98/24/EC (CAD); and
(b) the Explosive Atmospheres Directive 99/92/EC (ATEX).

16 The primary purpose of DSEAR is to protect workers and others who may be at risk from dangerous substances that can cause a fire, explosion or similar energy-releasing event, such as a runaway exothermic reaction.

DSEAR and other legislation on dangerous substances or hazardous activities

17 Subject to regulation 3(1), risks from dangerous substances or explosive atmospheres may also be subject to other specific legislation dealing with, for example:

(a) risks from major hazard activities, covered by the Control of Major Accident Hazards Regulations 1999 (SI 1999/743) (COMAH);
(b) the prevention of fires and explosions offshore through the Offshore Installations (Prevention of Fire and Explosion, and Emergency Response) Regulations 1995 (SI 1995/743) (PFEER) or the Offshore Installations and Pipeline Works (Management and Administration) Regulations 1995 (SI 1995/738); and
(c) the transport of dangerous goods, covered by specific carriage legislation, such as the Carriage of Dangerous Goods and Use of Transportable Pressure Equipment Regulations 2007 (CDG). (These have been superseded by the Carriage of Dangerous Goods and Use of Transportable Pressure Equipment Regulations 2009 (SI 2009/1348). Reference to the amendment has yet to be reflected formally in DSEAR but section 17(2)(a) of the Interpretation Act 1978 means that reference to the earlier Regulations includes the later Regulations.)

18 The above list is not exclusive and other legislation includes provisions on preventing fires and explosions for example in quarries, mines and harbour areas. In many cases, compliance with the specific requirements of such other legislation will go a long way towards meeting the requirements of DSEAR.

Relationship with other health and safety legislation

19 The duties in DSEAR apply alongside the HSW Act, other regulations made under the Act, and legislation on fire precautions and within a wider legislative context. See Appendix 2 for more information. Certain significant regulations are highlighted below.

General fire safety legislation

20 General fire safety requirements in the workplace in England and Wales are applied through the Regulatory Reform (Fire Safety) Order 2005 (SI 2005/1541). The Scottish equivalents are the Fire (Scotland) Act 2005 (2005 (asp 5)) and the Fire Safety (Scotland) Regulations 2006 (SSI 2006/456). The legislation consolidated and revoked legislation that previously covered general fire safety, including the requirement for fire certification. Current legislation requires the employer to carry out a risk assessment to determine the general fire safety requirements for their workplace and implement those requirements, including maintaining a general fire safety management plan.

21 General fire safety legislation is normally enforced by the local fire and rescue authority or in Scotland, the Scottish Fire and Rescue Service. However, there are other enforcing authorities for certain specific workplaces/worksites. HSE is the enforcing authority for general fire safety in shipbuilding and on most construction sites. At the time of publication, HSE – via the Office for Nuclear Regulation (ONR) – is the enforcing authority for fire safety in licensed nuclear premises.

22 Any general fire safety provision which could be imposed by regulations 1–6, 8, 9 and 11 of DSEAR is covered instead under this general fire safety legislation, with enforcement responsibility falling to the relevant authority, depending on the activity at the premises.

The Equipment and Protective Systems Intended for Use in Potentially Explosive Atmospheres Regulations 1996

23 The Equipment and Protective Systems Intended for Use in Potentially Explosive Atmospheres Regulations 1996 (SI 1996/192) (EPS) implement Directive 94/9/EC (the ATEX Product Directive). EPS applies to both electrical and mechanical equipment and protective systems intended for use in potentially explosive atmospheres.

24 New equipment etc supplied for use in places where an explosive atmosphere may occur must meet the requirements of EPS whatever its source. Second-hand equipment and equipment already on the shelf (for example, as spares) brought into use after 1 July 2003 also has to meet the requirements of EPS.

25 Regulation 7(6) of DSEAR made provisions in regulation 17 for transitional delays before certain requirements of regulation 7 and Schedule 3 (in relation to equipment in hazardous areas) became entirely active. All of the transitional delay periods have now expired but guidance on regulation 17 can now be found under regulation 7.

Environmental issues

26 DSEAR deals only with risks to people from dangerous substances but such substances could also harm the environment during disposal or in the event of a spill. In undertaking any risk assessment, or developing emergency arrangements, the potential for environmental harm should also be considered. Safety and environmental risks will need to be balanced and judgements will need to be made when considering substitution or other risk control/mitigation measures. Further guidance on environmental considerations is available from:

- the Environment Agency in England www.environment-agency.gov.uk/business/sectors/wastemanagement.aspx
- the Scottish Environment Protection Agency (SEPA) www.sepa.org.uk/waste.aspx
- Natural Resources Wales Cyfoeth Naturiol Cymru http://naturalresourceswales.gov.uk

Enforcement arrangements

27 DSEAR is enforced by HSE or local authority inspectors in accordance with the Health and Safety (Enforcing Authority) Regulations 1998 (SI 1998/494) except at commercial premises holding petroleum licences. At these premises the regulations are enforced by petroleum licensing authorities in respect of any activities related to refuelling means of transport such as motor vehicles.

28 At most workplaces, fire and rescue authorities will enforce equivalent provisions to DSEAR that relate to general fire safety.

Consulting employees and safety representatives

29 Proper consultation with those who know precisely how the work is done, including identifying the implications of any short cuts, is crucial and helps to build a culture of awareness of health and safety.

30 Employers must consult safety representatives appointed by recognised trades unions under the Safety Representatives and Safety Committees Regulations 1977 (SI 1977/500). Employees who are not covered by such representatives must be consulted either directly or indirectly, through elected representatives of employee safety under the Health and Safety (Consultation with Employees) Regulations 1996 (SI 1996/1513).

The Regulations

31 Regulations 1 to 4 deal with preliminary issues, ie the date of entry into force of the Regulations, scope and definitions.

32 The DSEAR Regulations reproduced in this ACOP include amendments made since 2002 up to mid-2013 only. You should ensure that the most recent version of any regulation is consulted. You can search for changes to legislation at www.legislation.co.uk.

Regulation 1 Citation and commencement

Summary of regulation 1

All parts of the regulations are now in force.

Regulation 1

These Regulations may be cited as the Dangerous Substances and Explosive Atmospheres Regulations 2002 and shall come into force –

(a) as respects all regulations except for regulations 5(4)(c), 7, 11, 15(2), 16(2) and 17(1) to (3) on 9th December 2002;
(b) as respects regulations 15(2) and 16(2) on 5th May 2003; and
(c) as respects regulations 5(4)(c), 7, 11 and 17(1) to (3) on 30th June 2003.

Regulation 2 Interpretation

Summary of regulation 2

Use the definitions listed here to check the exact meaning of terms used, eg a 'dangerous substance'.

Regulation 2

In these Regulations –

"approved classification and labelling guide" means the "Approved Guide to the Classification and Labelling of Dangerous Substances and Dangerous Preparations" (5th edition)[a] approved by the Health and Safety Commission on 16th April 2002;

"the CHIP Regulations" means the Chemicals (Hazard Information and Packaging for Supply) Regulations 2009;

"dangerous substance" means –

(a) a substance or preparation which meets the criteria in the approved classification and labelling guide for classification as a substance or preparation which is explosive, oxidising, extremely flammable, highly flammable or flammable, whether or not that substance or preparation is classified under the CHIP Regulations;

Regulation 2

(b) a substance or preparation which because of its physico-chemical or chemical properties and the way it is used or is present at the workplace creates a risk, not being a substance or preparation falling within subparagraph (a) above; or

(c) any dust, whether in the form of solid particles or fibrous materials or otherwise, which can form an explosive mixture with air or an explosive atmosphere, not being a substance or preparation falling within subparagraphs (a) or (b) above;

"explosive atmosphere" means a mixture, under atmospheric conditions, of air and one or more dangerous substances in the form of gases, vapours, mists or dusts in which, after ignition has occurred, combustion spreads to the entire unburned mixture;

"hazard" means the physico-chemical or chemical property of a dangerous substance which has the potential to give rise to fire, explosion, or other events which can result in harmful physical effects of a kind similar to those which can be caused by fire or explosion, affecting the safety of a person, and references in these Regulations to "hazardous" shall be construed accordingly;

"offshore installation" has the same meaning as it is given by regulation 3 of the Offshore Installations and Pipeline Works (Management and Administration) Regulations 1995 insofar as that regulation extends to mineral extracting industries within the scope of Article 2(a) of Council Directive 92/91/EEC concerning the minimum requirements for improving the safety and health protection of workers in the mineral-extracting industries through drilling;

"personal protective equipment" means all equipment which is intended to be worn or held by a person at work and which protects that person against one or more risks to his safety, and any addition or accessory designed to meet that objective;

"preparation" means a mixture or solution of two or more substances;

"public road" means (in England and Wales) a highway maintainable at public expense within the meaning of section 329 of the Highways Act 1980 and (in Scotland) a public road within the meaning assigned to that term by section 151 of the Roads (Scotland) Act 1984;

"risk" means the likelihood of a person's safety being affected by harmful physical effects being caused to him from fire, explosion or other events arising from the hazardous properties of a dangerous substance in connection with work and also the extent of that harm;

"risk assessment" means the assessment of risks required by regulation 5(1);

"safety data sheet" means a safety data sheet within the meaning of Regulation (EC) No 1907/2006 of the European Parliament and of the Council concerning the Registration, Evaluation, Authorisation and Restriction of Chemicals;

"substance" means any natural or artificial substance whether in solid or liquid form or in the form of a gas or vapour;

"workplace" means any premises or part of premises used for or in connection with work, and includes –

(a) any place within the premises to which an employee has access while at work; and

Regulation 2

(b) any room, lobby, corridor, staircase, road or other place –
(i) used as a means of access to or egress from that place of work; or,
(ii) where facilities are provided for use in connection with that place of work;

other than a public road; and

"work processes" means all technical aspects of work involving dangerous substances and includes –

(a) appropriate technical means of supervision,
(b) connecting devices,
(c) control and protection systems,
(d) engineering controls and solutions,
(e) equipment,
(f) materials,
(g) machinery,
(h) plant,
(i) protective systems, and
(j) warning and other communication systems.

(a) The Approved Guide to the Classification and Labelling of Dangerous Substances and Dangerous Preparations (Sixth edition) 2009 is the most current version.

Guidance 2

33 In this regulation:

(a) 'Other events' under the regulation 2 definition of 'hazard' of the dangerous substance include exothermic runaway reactions.
(b) 'Connecting devices' under the regulation 2 definition of 'work processes' are the means by which discrete parts of the process are interconnected, eg conveyor belts, trunking, pipework etc.
(c) 'Dangerous substance' includes any substance or preparation which, because of its properties or the way it is used, could cause harm to people from fires and explosions. Dangerous substances include those with potential energy releasing events similar to fire and explosion such as exothermic reactions. Examples include: petrol; liquefied petroleum gas (LPG); paints; varnishes; solvents; and dusts which when mixed with air could cause an explosive atmosphere (eg dusts from milling and sanding operations).

34 Substances and preparations are potentially within the scope of the Regulations – whether in solid, liquid or gaseous form. This includes substances that are naturally occurring or produced in a chemical or manufacturing process. Substances are also included if they are produced by a work activity, for example, intermediates in a chemical process, waste products of any kind, or substances produced in accident conditions, eg in a runaway chemical reaction.

35 The Regulations apply wherever a dangerous substance is, or is liable to be, used or present in connection with a work activity carried out by an employer.

36 Although the chemical and petroleum industries will by their very nature store, use and process the majority of dangerous substances, most other sectors (such as manufacturing, food, retailing etc) will also have dangerous substances present but possibly in small quantities.

Guidance 2

Key terms explained

37 The following provides definition of some terms used within this ACOP:

(a) 'control' is used to describe steps taken or provisions put in place to reduce the likelihood of a fire, explosion, or similar event happening. Control measures should be considered before considering mitigation;
(b) 'mitigation' is used to describe what steps are taken to minimise the consequences during and after the occurrence of a fire, explosion or similar event;
(c) certain combustible dusts are explosible. This means that when dispersed in air in the right dust/air mixture they can cause an explosion if ignited. For simplicity the term 'explosive' is used in this ACOP to describe these dusts. In this context the term does not mean materials which are in the class of intentionally explosive materials (UN class 1);
(d) reducing risk as low as reasonably practicable (ALARP) – detailed guidance on the principles of ALARP may be found on the HSE website at www.hse.gov.uk/risk/theory/alarp.htm.

Approved classification and labelling

38 See Appendix 3 which explains changes in classification arrangements.

Explosive atmospheres

39 This definition sets out the criteria to determine whether an atmosphere is explosive within the scope of the Regulations. The definition is particularly important in deciding when certain requirements in DSEAR will apply, particularly regulations 7 and 11.

40 The following three points are consistent with the guidance on EPS, where 'explosive atmosphere' is similarly defined. This is relevant to regulation 7 of DSEAR and the selection of equipment for use in explosive atmospheres. For the purposes of DSEAR the following elements must all be present for an explosive atmosphere which is subject to regulations 5(4)(c), 7 and 11 to form:

(a) **atmospheric conditions** – for the purposes of standardisation, normal atmospheric conditions are defined as -20 °C to 40 °C, and 0.8 to 1.1 bar;
(b) **mixtures of air and dangerous substances** – dangerous substances or mixtures of such substances, that are explosive with an oxidant other than air, for example pure oxygen or chlorine, are outside the scope of the definition of explosive atmosphere. The provisions of regulations 5(4)(c), 7 and 11 do not apply although other requirements in DSEAR may do;
(c) **combustion** – the definition of explosive atmosphere in regulation 2 is intended to make clear that where it can be ensured that the gas or dust is present in a concentration below the lower explosive limit, the atmosphere is not explosive, and regulations 5(4)(c), 7 and 11 do not apply. In this guidance, in the terms 'flammable limit' and 'explosive limit' (as used for example in 'lower flammable limit'), the words 'flammable' and 'explosive' are intended to have the same meaning and are interchangeable. For simplicity, the term 'lower explosive limit' (LEL) is used throughout this document.

Guidance 2

Hazard

41 In common usage, a hazard is anything with the potential to cause harm in any way, but a more restricted meaning is specified within DSEAR when considering dangerous substances and preparations. In DSEAR the term 'hazard' is confined to the properties of a substance that can potentially lead to fire or explosion or other similar energetic effects which could affect a person's safety.

42 The Regulations are concerned with dangerous substances that can create harmful physical effects. However, DSEAR does not apply to all possible physical effects, such as crushing injuries resulting, for example, from the storage and handling of heavy containers of dangerous substances and preparations. The Regulations are intended to cover the following harmful physical effects caused directly or indirectly by fires, explosions and other similar events including:

(a) thermal radiation (burns caused by radiating heat);
(b) thermal injury (burning substances on the skin);
(c) over-pressure (blast injuries);
(d) smoke, fire gases, unintended releases (asphyxiation).

Other events

43 Harmful physical effects can also be caused by other energetic events such as runaway exothermic reactions or decompositions of unstable substances, eg decomposition of peroxides. These words are included in the definition of hazard to ensure that not only fires and explosions are covered, but also other similar energetic events.

Physico-chemical or chemical property

44 Ultimately the hazard is created by the physico-chemical and chemical properties of the substance or preparation and the way it is used or present.

45 Examples of relevant physical properties include:

(a) boiling point, flashpoint, auto-ignition temperature;
(b) flammability, vapour pressure, thermal sensitivity;
(c) mechanical sensitivity and oxidising properties.

46 Relevant chemical properties would include:

(a) reactivity, heat of reaction;
(b) self-acceleration and decomposition temperature.

47 Other properties of substances relating to radioactivity, toxicity and ecotoxicity are not within the scope of the Regulations.

48 Test methods that can be used to determine physico-chemical properties are detailed in the Test Methods Regulation (TMR) (EC) No 440/2008 or the UN Manual of Tests and Criteria. Further information on TMR can be found on the European Chemicals Agency website at http://echa.europa.eu.

49 For materials that could be dispersed in air to give rise to a risk of a dust explosion, the consequences and magnitude of this are significantly influenced by the composition and nature of the material, including its particle size. A number of European Standards for test methods to help determine the risk and the precautionary measures required are available (BS EN 14034 Parts 1–4[3] and BS EN 13821[4]).

Guidance 2

Safety data sheets

50 Safety data sheets (SDSs) are no longer covered by the Chemicals (Hazard Information and Packaging for Supply) Regulations 2009 (SI 2009/716) (CHIP). The legal requirements to provide an SDS have been transferred to the direct-acting Registration, Evaluation, Authorisation and Restricting of Chemicals Regulations (EC 1907/2006) (REACH Regulation).

Workplace

51 The definition of 'workplace' is based on that used in the Workplace (Health, Safety and Welfare) Regulations 1992 (SI 1992/3004). It is however wider in scope as it also includes areas in private dwellings where work is carried out.

52 The term 'premises', used in the definition of 'workplace', means any place whether or not there is a structure at that place. It includes (subject to the disapplications in regulation 3) vehicles, vessels, any land-based or offshore installations, movable areas to which employees have access while at work and their means of access to and egress from, the workplace. Common parts of shared buildings, private roads and paths on industrial estates and business parks are therefore included.

53 Public roads which are used to get to or from the workplace are not included in the definition. However, in some circumstances, a public road may itself become the workplace, and if dangerous substances are used or produced during the work activity concerned, these Regulations may apply, for example during road repairing or work on utilities.

Work processes

54 This includes the use of the 'hardware' aspects of work involving dangerous substances that are of a technical nature, eg gas detectors, flameproof enclosures, regulating devices. It excludes systems of work, eg management and non-technical supervisory arrangements, but does include appropriate technical measures for supervision. Work processes can involve a range of activities, eg storage, manufacture, disposal, cleaning and some forms of energy generation.

Technical supervision measures

55 Technical means of supervision are those technical measures that are required by the risk assessment to prevent employees from working in or entering an explosive atmosphere. They will include monitoring and interlock devices that are designed to stop a process or alert an employee in the event of a mechanical fault, procedural error or foreseeable process deviation that could result in the formation of a hazardous explosive atmosphere. Examples of such technical measures include:

(a) gas monitoring and alarm systems for employees working in confined spaces where flammable vapours may be present or likely to be released by the activity;
(b) interlocks on coating operations that would stop the application of flammable coating products if the associated mechanical exhaust ventilation is interrupted.

56 Technical means of supervision may be used as a separate supervisory function or as part of a wider management system for controlling risks.

Regulation 3 Application

Summary of regulation 3

To avoid overlaps with other more specific legislation regulation 3 disapplies some named work activities from some individual DSEAR regulations. These work activities include maritime, medical, gas appliances, explosives manufacture, mines and quarries, and offshore installations.

Regulation 3

(1) These Regulations, apart from regulations 15, 16 and 17(4) to (5), shall not apply to the master or crew of a ship or to the employer of such persons in respect of the normal ship-board activities of a ship's crew which are carried out solely by the crew under the direction of the master and, for the purposes of this paragraph –

(a) "ship" includes every description of vessel used in navigation, other than a ship forming part of Her Majesty's Navy or an offshore installation; and
(b) the reference to the normal ship-board activities of a ship's crew includes –
(i) the construction, reconstruction or conversion of a ship outside, but not inside, Great Britain; and
(ii) the repair of a ship save repair when carried out in dry dock.

(2) Regulations 5(4)(c), 7 and 11 shall not apply to –

(a) areas used directly for and during the medical treatment of patients;
(b) the use of gas appliances burning gaseous fuel (that is to say, any fuel which is in a gaseous state at a temperature of 15°C under a pressure of 1 bar) which –

(i) are used for cooking, heating, hot water production, refrigeration, lighting or washing; and
(ii) have, where applicable, a normal water temperature not exceeding 105°C;

including forced draught burners and heating bodies to be equipped with such burners but not including an appliance specifically designed for use in an industrial process carried out on industrial premises;

(c) gas fittings within the meaning of the Gas Safety (Installation and Use) Regulations 1998 located in domestic premises, not being gas appliances falling within subparagraph (b);
(d) the manufacture, handling, use, storage and transport of explosives or chemically unstable substances;
(e) any activity at a mine within the meaning of section 180 of the Mines and Quarries Act 1954 carried out for the purposes of the mine;
(f) any activity at a quarry within the meaning of regulation 3 of the Quarries Regulations 1999 carried out for the purposes of the quarry;
(g) any activity at a borehole site within the meaning of regulation 2(1) of the Borehole Sites and Operations Regulations 1995 carried out for the purposes of the borehole site;
(h) any activity at an offshore installation carried out for the purposes of the offshore installation; and
(i) the use of means of transport by land, water or air which is regulated by international agreements and the European Community Directives giving effect to them insofar as they fall within the disapplication in Article 1.2.(e) of Council Directive 99/92/EC on minimum requirements for

Regulation 3

improving the safety and health protection of workers potentially at risk from explosive atmospheres, except for any means of transport intended for use in a potentially explosive atmosphere.

(3) Regulations 5(2)(f), (g), (h) and (i), 6(4)(d), 6(5)(b) and (e) and 8(1)(d) and (e) and the requirements of paragraphs 5 and 6 of Schedule 1 shall not apply to any activity at an offshore installation carried out for the purposes of the offshore installation.

Guidance 3

Maritime activities

57 The Regulations do not apply to most activities on ships carried out solely by a ship's crew involving dangerous substances on ships under the direction of the ship's master. For example, use of flammable paints for maintenance purposes. Similarly when the ship is tied up in a port or harbour, minor 'running repairs' involving dangerous substances (eg repairs involving flammable adhesives) carried out solely by the crew under the direction of the ship's master are not covered by DSEAR. The crew's safety is subject to maritime legislation enforced by the Maritime and Coastguard Agency, who also deal with navigation and other operational matters.

58 However, DSEAR does apply when a ship is in a British port and 'shoreside' workers and the ship's crew work together:

(a) in port/dock operations;
(b) in carrying out construction, reconstruction, or conversion repairs to the ship in dry dock.

59 Any work on offshore installations is not regarded as normal shipboard activities (and DSEAR applies). But where only the offshore staff are involved the operations are only excluded from DSEAR regulations 5(2)(f), (g), (h) and (i), 6(5)(b) and (e), 8(1)(d) and (e) and paragraphs 5 and 6 of Schedule 1. (There are specific requirements in offshore legislation.)

60 Regulation 3(1) also enables the provisions of regulations 15, 16, 17(4) and 17(5) to be applied in full to water transport and shipping activities where they may be subject to the legislation referred to in those regulations.

Exclusions from regulations 5(4)(c), 7 and 11

61 Regulation 3(2) lists a number of areas and activities to which regulations 5(4)(c) (recording information where an explosive atmosphere may occur at the workplace), 7 (places where explosive atmospheres may occur) and 11 (duty of co-ordination) do not apply.

Areas used directly for and during the medical treatment of patients

62 Only specific areas where medical treatment takes place such as treatment rooms and operating theatres are excluded. Other areas in hospitals or surgeries where treatment does not take place, such as waiting rooms, corridors, boiler rooms, laundries, fitters' workshops, and treatment areas closed for repairs or refurbishment, are not excluded.

Guidance 3

The use of gas appliances burning gaseous fuels

63 Appliances excluded from DSEAR are those which burn gaseous fuels used for cooking, heating, water heating, refrigeration, lighting and washing with, where applicable, a normal water temperature not exceeding 105 °C. Appliances designed for use in industrial processes on industrial premises and appliances that heat water to a temperature greater than 105 °C are covered by DSEAR. Points to note:

(a) 'Gaseous fuels' means any fuel which is a gas at a temperature of 15 °C and a pressure of 1 bar (normal atmospheric pressure).
(b) 'Use of appliances' does not include installation and maintenance.
(c) The pipework supplying an appliance is not included in the term 'use of the appliance' although the section between (and adjacent to) the appliance and the isolation valve may be.
(d) The use of gas appliances is covered by the Gas Appliances (Safety) Regulations 1995 (SI 1995/1629).

Gas fittings

64 The requirements of regulations 5(4)(c), 7 and 11 do not apply to work in domestic premises involving the installation and use of gas fittings as defined in the Gas Safety (Installation and Use) Regulations 1998 (SI 1998/2451).

The manufacture, handling, use, storage and transport of explosives or chemically unstable substances

65 Regulations 5(4)(c), 7 and 11 (recording information, places where explosive atmospheres may occur and duty of co-ordination) do not apply to activities involving the manufacture, handling, use and storage of explosives and chemically unstable substances, or to their transport by road, rail, water and air.

66 In DSEAR, 'explosives' means those substances (and articles) that meet the interpretation of 'explosive' given in the Manufacture and Storage of Explosives Regulations 2005 (SI 2005/1082).

67 In DSEAR, 'chemically unstable substances' means substances or preparations which in foreseeable conditions and without the need for air, either individually or when mixed with another substance, can undergo a self-sustaining chemical reaction. This reaction can produce heat and/or gas at such a temperature or rate that it poses a risk to safety from blast, thermal or projectile effects. They include many organic peroxides[5] and various blowing agents of the 'azo' type. For regulations 5(4)(c), 7 and 11 to apply, the explosive atmosphere must be formed under normal atmospheric conditions, ie as a mixture in air.

68 Some substances have both the properties of a flammable gas, vapour, dust etc that can form an explosive atmosphere with air under normal atmospheric conditions **and** can also explode in the absence of air. Examples include acetylene, methyl acetylene and ethylene oxide. Regulations 5(4)(c), 7 and 11 therefore apply to these substances in relation to the possible formation of explosive atmospheres in the presence of air. The three DSEAR regulations do not apply in relation to their explosive/chemically unstable properties. Further information on the additional legal and practical controls applying to acetylene can be found on the HSE website at www.hse.gov.uk/fireandexplosion/acetylene.htm.

Guidance 3

Mineral extracting industries

69 Regulations 5(4)(c), 7 and 11 do not apply to activities at mines (within the meaning of the Mines and Quarries Act 1954), quarries (within the meaning of the Quarries Regulations 1999 (SI 1999/2024)) or borehole sites (Borehole Sites and Operations Regulations 1995 (SI 1995/2038)) where these are carried out for the specific purpose of the extraction. Activities at an offshore installation carried out for the purposes of the installation are also excluded from these regulations.

70 Specifically the requirements for visual or audible warnings (regulations 5(2)(f), (g), (h) and (i), 6(4)(d), 6(5)(b) and (e)) and escape facilities in emergencies (regulations 8(1)(d) and (e)) and appropriate systems of work (paragraphs 5 and 6 of Schedule 1) also do not apply to activities at offshore installations because similar requirements exist in other specific offshore legislation.

Use of means of transport

71 Regulation 3(2)(i) means that the provisions in regulations 5(4)(c), 7 and 11 (outlined in paragraph 62) do not apply to means of transport of dangerous goods covered by EU directives and international agreements (ADR, RID etc) as implemented nationally, unless the transport (eg a vehicle) is intended to be used in a potentially explosive atmosphere. Even if these three sections are disapplied because other requirements cover the issues during transit, the rest of DSEAR will still apply to any workplace where a dangerous substance is liable to be present and where there may be a risk of an explosive atmosphere, eg from a spillage. Therefore where no international transport agreements are involved, regulations 5(4)(c), 7 and 11 will not be disapplied.

72 Loading/unloading operations which take place in an area of a site where there is normally a risk of a potentially explosive atmosphere are not excluded and the requirements of regulations 5(4)(c), 7 and 11 apply to the means of transport. If an explosive atmosphere is likely to occur solely as a result of a loading/unloading operation itself then the operation would be excluded only from the three specific sections mentioned – the rest of DSEAR would apply. For example, a visiting road tanker transporting a dangerous substance would not bring application of hazardous area classification to the site it was visiting, where the site is not initially considered to have a potentially explosive atmosphere because of its location with respect to the storage facility.

73 The more specific provisions of the legislation on the transport of dangerous goods apply as normal to ensure safety during transport, loading and unloading.

74 Any transport of dangerous goods outside the scope of international agreements etc is not excluded and is fully subject to regulations 5(4)(c), 7 and 11. This includes vehicles that remain on the employer's premises, such as forklift trucks, working/loading/unloading in potentially explosive atmospheres.

Regulation 4 Duties under these Regulations

Summary of regulation 4

This explains the scope of the Regulations as they apply to the duties of employers and the self-employed who are included within the term 'employer'. Duties to employees and others are outlined including provision of protective equipment and instruction.

Regulation 4

(1) Where a duty is placed by these Regulations on an employer in respect of his employees, he shall, so far as is reasonably practicable, be under a like duty in respect of any other person, whether at work or not, who may be affected by the work carried on by the employer, except that –

(a) the duties of the employer under regulations 6(5)(f) and 7(5) (which relate, respectively, to the provision of suitable personal protective equipment and the provision of appropriate work clothing) shall not extend to persons who are not his employees; and

(b) the duties of the employer under regulations 8 and 9 (which relate, respectively, to dealing with accidents and to provision of information, instruction and training) shall not extend to persons who are not his employees, unless those persons are at the workplace where the work is being carried on and subject to the following, namely, that, in relation to the application of regulation 9 to such persons, regulation 9 shall apply to the extent that is required by the nature and the degree of the risk.

(2) These Regulations shall apply to a self-employed person as they apply to an employer and an employee and as if that self-employed person were both an employer and employee.

Guidance 4

75 These Regulations place specific duties on employers to assess and control the risks from dangerous substances. For the purposes of these Regulations, employers include contractors, subcontractors and self-employed people. The duties under these Regulations apply to a self-employed person as if they were both an employer (as long as their activities could affect others) and an employee.

76 In addition to their own employees, employers are required under DSEAR (and by section 3 of the HSW Act) to take account of the risks to people who are not employed by them but who may be at risk from the use or presence of a dangerous substance. This includes anyone who could potentially be at risk including employees working for other employers, visitors to the site, anyone such as members of the public occupying premises or space nearby.

77 However, under regulation 8 (arrangements to deal with accidents, incidents and emergencies) and regulation 9 (information, instruction and training), employers only have duties to people other than their employees when those people are at the site of the work activity. In addition, employers are not required by law to provide people other than their employees with appropriate personal protective equipment or work clothing. Information, instruction and training need only be provided to non-employees to the extent required by the nature and degree of the risk.

78 Where employees of one employer work at another employer's premises, both employers have duties under the Regulations. Each employer has duties to their own and the other employer's employees. The employers should co-operate and collaborate to ensure that all the duties imposed by these Regulations are fulfilled. Arrangements for this should be agreed between them, but each employer must satisfy themselves that any arrangements adopted are adequate.

Guidance 4

79 Where an explosive atmosphere may occur, the employer responsible should co-ordinate the implementation of any explosion protection measures. (See regulation 11.)

Regulation 5 Risk assessment

Summary of regulation 5

This regulation requires all employers and the self-employed to assess all potential risks to employees and others whose safety may be affected by the use or presence of a dangerous substance at the workplace.

Employers should check that measures are in place before beginning any new work activity or processes and recheck arrangements periodically.

If five or more people are employed, a written record of the significant findings is required. This can either be a separate record or integrated into the record made of overall assessment under the Management of Health and Safety at Work Regulations 1999 (SI 1999/3242) (the Management Regulations) (see Appendix 2) or the fire safety regulations (see paragraph 20).

Regulation 5

(1) Where a dangerous substance is or is liable to be present at the workplace, the employer shall make a suitable and sufficient assessment of the risks to his employees which arise from that substance.

(2) The risk assessment shall include consideration of –

(a) the hazardous properties of the substance;
(b) information on safety provided by the supplier, including information contained in any relevant safety data sheet;
(c) the circumstances of the work including –
  - *(i) the work processes and substances used and their possible interactions;*
  - *(ii) the amount of the substance involved;*
  - *(iii) where the work will involve more than one dangerous substance, the risk presented by such substances in combination; and*
  - *(iv) the arrangements for the safe handling, storage and transport of dangerous substances and of waste containing dangerous substances;*

(d) activities, such as maintenance, where there is the potential for a high level of risk;
(e) the effect of measures which have been or will be taken pursuant to these Regulations;
(f) the likelihood that an explosive atmosphere will occur and its persistence;
(g) the likelihood that ignition sources, including electrostatic discharges, will be present and become active and effective;
(h) the scale of the anticipated effects of a fire or an explosion;
(i) any places which are or can be connected via openings to places in which explosive atmospheres may occur; and
(j) such additional safety information as the employer may need in order to complete the risk assessment.

(3) The risk assessment shall be reviewed by the employer regularly so as to keep it up to date and particularly if –

(a) there is reason to suspect that the risk assessment is no longer valid; or

Regulation 5

(b) there has been a significant change in the matters to which the risk assessment relates including when the workplace, work processes, or organisation of the work undergoes significant changes, extensions or conversions;

and where, as a result of the review, changes to the risk assessment are required, those changes shall be made.

(4) Where the employer employs five or more employees, the employer shall record the significant findings of the risk assessment as soon as is practicable after that assessment is made, including in particular –

(a) the measures which have been or will be taken by him pursuant to these Regulations;
(b) sufficient information to show that the workplace and work processes are designed, operated and maintained with due regard for safety and that, in accordance with the Provision and Use of Work Equipment Regulations 1998, adequate arrangements have been made for the safe use of work equipment; and
(c) where an explosive atmosphere may occur at the workplace and subject to the transitional provisions in regulation 17(1) to (3), sufficient information to show –
  (i) those places which have been classified into zones pursuant to regulation 7(1);
  (ii) equipment which is required for, or helps to ensure, the safe operation of equipment located in places classified as hazardous pursuant to regulation 7(1);
  (iii) that any verification of overall explosion safety required by regulation 7(4) has been carried out; and
  (iv) the aim of any co-ordination required by regulation 11 and the measures and procedures for implementing it.

(5) No new work activity involving a dangerous substance shall commence unless –

(a) an assessment has been made; and
(b) the measures required by these Regulations have been implemented.

ACOP 5

Risk assessment

80 During the risk assessment employers should identify safety risks arising out of or in connection with work or the conduct of their undertaking that relate to dangerous substances and should identify who is at risk. The findings should then be used to take practical action (regulations 6 and 7) to eliminate or reduce the risk. The two aspects (identification of risk and resulting action) are interlinked and so regulations 5, 6 and 7 should be considered together. For the risk assessment to meet the legal requirement to be suitable and sufficient, employers should cover all points in regulation 5(2) to 5(3) inclusive as a minimum and should:

(a) include a determination of the hazardous properties of the dangerous substance(s);
(b) identify those different groups of workers and people who may be harmed and the likelihood and severity of the consequences;
(c) consider any employees who may be at increased risk because of lack of awareness, eg inexperienced trainees and those under 18;
(d) consider others including workers of another employer in the workplace

ACOP 5

or nearby, members of the public and other visitors, both on and off site;

(e) satisfy themselves that where a 'model' risk assessment is being used from plants elsewhere using similar processes, in each case, the model:
- **(i) reflects the core hazards;**
- **(ii) is adapted to the detail of the particular situation;**
- **(iii) is appropriate to the type of work.**

Guidance 5

81 The risk assessment required by regulation 5 involves identification and careful examination of any dangerous substances present or liable to be present in the workplace and consideration of how fire, explosion and similar events might harm employees and any other people affected by the work concerned. Its purpose is to enable employers to decide what they must do to eliminate or reduce the risks from dangerous substances, so far as reasonably practicable.

82 The assessment (including the recording of significant findings) enables employers to demonstrate to themselves and to others who may have an interest, eg inspectors, employees' representatives (including safety and trade union representatives) etc, that they have followed a structured and thorough approach in considering the risks to the safety of employees and the control measures that are needed. Whoever carries out the assessment should be competent to do so.

83 Employers should carry out hazardous area classification (see regulation 7) as an integral part of the risk assessment to identify places where controls over ignition sources are needed and those places where they are not. Schedule 2 of the Regulations (located with regulation 7) sets out definitions of the zones to be used when classifying hazardous areas.

84 The DSEAR risk assessment may be carried out as part of that required by the Management Regulations. Where this is not done, a separate DSEAR risk assessment will be required.

85 Many factors influence the risks from a fire involving dangerous substances. In particular, employers should consider:

(a) whether a fire could lead to an explosion;
(b) how fast a fire might grow;
(c) what other materials might be rapidly evolved;
(d) any dangers from smoke and toxic gases given off;
(e) and whether those in the vicinity would be able to escape.

86 Further information on how to undertake a suitable and sufficient risk assessment is available on HSE's risk management webpages at www.hse.gov.uk/risk.

87 The risk assessment could, where appropriate, be completed alongside goal-orientated risk assessments, such as the BS EN 61508-1[6] or BS EN 61511-2[7] sector standards used by the process industry, providing that risks from fire, explosion and other events arising from dangerous substances are addressed, including the requirements specified by DSEAR.

88 Employers' risk assessments (and the record of the assessment where one is required) should take account of the presence of dangerous substances on the effect of the general fire precautions/safety requirements. The general fire safety legislation follows the same approach as DSEAR in requiring the employer to carry out a risk assessment to identify the risks to people from a fire at the employer's premises and the practical actions taken to eliminate or reduce the risk.

Guidance 5

89 General fire precautions include provision of:

(a) adequate and appropriate means of detection and giving warning in case of fire;
(b) adequate means of escape;
(c) suitable means of fighting fire;
(d) specifying the action to be taken in the event of fire; and
(e) appropriate and adequate training of staff in company fire safety procedures.

90 An explanation of how to comply with the law relating to general fire safety requirements and how to carry out a fire risk assessment can be found:

(a) for England and Wales, in *Fire safety in the workplace* https://www.gov.uk/workplace-fire-safety-your-responsibilities;
(b) for Scotland, in *Fire Law – Are you aware of your responsibilities – Fire Safety Risk Assessment* www.scotland.gov.uk/Topics/Justice/public-safety/Fire-Rescue/FireLaw;
(c) for construction sites, in *Fire safety in construction* HSG168.[8]

91 In most workplaces, the local fire and rescue authority enforces the general fire safety legislation which covers general fire safety precautions required in case of fire. In so far as they relate to general fire safety, this includes equivalent provision to regulations 1–6, 8, 9 and 11 of DSEAR. Other requirements for preventing and controlling fires under DSEAR (such as measures to prevent leaks of dangerous substances and avoiding sources of ignition) are enforced by HSE or the local authority, depending on the activity in the premises.

The hazardous properties of a substance

92 A suitable and sufficient risk assessment should give consideration to the overall risk presented by dangerous substances as well as assessing each factor individually.

93 Employers must identify any dangerous substances that may be present at the workplace and the hazards they present (eg their flammable or explosive properties). This includes substances which are:

(a) brought into the workplace and handled, stored and used for processing;
(b) produced or given off (eg as fumes, vapour, dust etc) by a process or activity, or as a result of an incident or accident;
(c) used for or arise from maintenance, cleaning and repair work; or
(d) produced as a by-product of any work or process (eg waste, residues, scrap materials etc);
(e) naturally occurring in the workplace (eg methane may be present in tunnelling and mining operations).

94 When considering information on the hazardous properties of dangerous substances, employers should identify any adverse conditions that should be avoided. These could include excessive heat, sunlight, exposure to air or moisture and contact with other incompatible substances.

95 Some substances are obviously hazardous to safety, other substances might be hazardous only under certain conditions. Flour dust can form an explosive atmosphere, and liquids such as diesel fuel can be raised above their flashpoint temperature by work activities and present a fire or explosion risk. When carrying out hazardous area classification, relevant properties of a dangerous substance include the boiling point and flashpoint of any flammable liquid, and whether any flammable gas or vapour that may be evolved is lighter or heavier than air.

Guidance 5

96 For dusts which are liable to form an explosive atmosphere, information will be needed on particle size and potential concentration in air.

97 A mass of solid combustible material as a heap or pile will burn relatively slowly owing to the limited surface area exposed to the oxygen of the air. A dust explosion involves the rapid combustion of dust particles that releases energy and usually occurs when dusts are dispersed in air, generating gaseous reaction products. A basic knowledge of the material's properties together with the avoidance of deposition – and in particular deposition on elevated surfaces – of dusts/powders which are capable of rapid explosive burning in air should minimise the risks of a dust explosion.

Information on hazardous properties

98 Useful information on the properties and hazards of dangerous substances may be provided by suppliers, eg in an SDS. This could include details such as flashpoints or explosive or chemical properties. Other information could relate to safe methods of using, storing and handling the substances. Suppliers are required to make SDSs available for the chemicals they supply and they are a source of information on the classification of the substance. Further information on the background to this is in Appendix 3.

99 The increased alignment in the EU CLP Regulation (classification, labelling and packaging) (no 1272/2008) towards a globally harmonised system of classification and labelling of substances and mixtures has led to a number of substances now meeting the criteria for classification as flammable. This is partly because the upper flashpoint for flammable liquid has been increased from 55 °C to 60 °C. The changes mean that for example, diesel, gas oil and light heating oils are now classified as flammable liquids.

100 However, many substances so classified may in fact not normally present a significant risk of fire as stored. Employers should adopt a proportionate approach in considering whether there are any justifiable further measures needed in addition to those widely used before this change, given that the risk itself has not changed.

101 Many dusts are not classified substances under CHIP and for these materials there is no legal requirement to provide an SDS. Suppliers should nevertheless be asked if they can supply any data relevant to assessing the fire and explosion risks. Where there is no SDS available in the case of intermediate products, mixtures or novel activities, perhaps under non-standard conditions, if databases such as the IFA GESTIS-Dust-Ex database available on the internet do not assist then further research and testing might be needed into the flammable/explosive hazards. This is also relevant under regulation 5(2)(j).

102 An assessment which made no reference to an SDS or similar information on hazardous properties researched in some other way would be unlikely to be considered suitable and sufficient especially if the risk was significant. However, some of the substances classified as flammable that have a high flashpoint may not give rise to a significant risk of fire unless processed at high pressure or at temperatures above their flashpoint. A detailed risk assessment for these substances on the way they are used or handled may indicate that fire risk reduction requirements can be relaxed to some degree.

ACOP 5

Circumstances of the work

103 When assessing and/or designing activities involving dangerous substances (such as work processes, process and storage plant and the workplace itself) all relevant factors must be taken into account, including:

(a) properties of the substances, including corrosivity, reactivity, volatility, flashpoint and electrical conductivity;
(b) quantities and storage methods, eg in bulk tanks or in containers;
(c) location relative to other features, including adjacent premises, site boundaries, occupied buildings, process areas, heat sources, fixed sources of ignition, other dangerous substances and vehicle thoroughfares;
(d) loading/unloading operations and frequency of deliveries.

104 The risk assessment should include consideration of whether work processes may give rise to flammable gases, vapours, mists or dusts in sufficient quantity to pose a risk of injury if ignited. The employer should consider the potential for the incident to escalate. Account should also be taken of possible accumulations of combustible dust which could be launched and dispersed into the air during an incident resulting in the formation of an explosive atmosphere.

105 When considering risks from releases of dangerous substances the following should be included:

(a) unavoidable releases, such as during LPG cylinder/aerosol filling;
(b) intentional releases, such as spray coating, solvent vapour emissions from drying ovens etc; and
(c) foreseeable releases, for example leaks from process equipment or storage containers or spills during dispensing.

Guidance 5

106 Adopt a methodical approach to considering the circumstances of the work activity, particularly the production processes. Consider the potential for and consequences of failures and/or errors or other foreseeable deviations from the way the work activity is carried out normally. The approach should be proportionate to the nature of the work activity and the risk it presents. Possible deviations and excursions from the norm are far more numerous for production processes than for a less dynamic storage activity. Guidance on safe storage is readily available (including some 'off-the-shelf' industry codes), whereas the wide variation possible in different production processes means that identifying potential failures during a process may need to be considered from first principles.

107 Employers need to consider potential hazards arising from equipment etc used in processing or handling dangerous substances. This includes equipment that may be brought into an area where dangerous substances are present as well as the equipment used in process operations themselves.

108 In taking account of the work activities involving dangerous substances when assessing risk, employers should include such activities as:

(a) loading and unloading operations (and the frequency of delivery/dispatch of dangerous substances);
(b) dispensing and decanting activities;
(c) movement of dangerous substances around the site; and
(d) how spillages and leaks are dealt with.

Guidance 5

109 The size of any potential release of dangerous substances is in part related to the amount of dangerous substances present. Industry-specific codes from various sources provide guidance on the quantities of various dangerous substances that can be stored, for example:

(a) Code of Practice 7 *Storage of Full and Empty LPG Cylinders and Cartridges*, produced by the UKLPG;[9]
(b) *Guidance for the storage of gas cylinders in the workplace* GN 2, produced by the British Compressed Gas Association.[10]

For further guidance on intermediate bulk container (IBC) storage, see

(c) joint Chemical Business Association (CBA) and Solvents Industry Association (SIA) guidance: *Guidance for the storage of liquids in intermediate bulk containers*.[11]
(d) the Energy Institute's Model Code of Safe Practice Part 15 *Area Classification Code for Installations Handling Flammable Fluids*;[12]
(e) HSE's guidance on storage and use of flammable liquids:
    (i) *The storage of flammable liquids in containers* HSG51;[13]
    (ii) *The storage of flammable liquids in tanks* HSG176;[14]
    (iii) *The safe use and handling of flammable liquids* HSG140.[15]

110 Some combinations of dangerous substances may react together to form an ignition source, or in combination may form an explosive atmosphere where singly this does not occur. Such possibilities should be considered in the risk assessment. For example:

(a) where substances are used together in a chemical process it will be necessary to carry out a chemical reaction hazard assessment to identify any adverse conditions that could give rise to hazardous heat and pressure effects;
(b) where different substances are stored, incompatible materials, such as self-reactive substances and flammable liquids, should not be kept together;
(c) where substances with different properties are processed together the fire properties of any resulting mixture will be different from the individual components. The properties of the mixture must be known to ensure, for example, any electrical equipment is suitable and any explosion relief provided is appropriate;
(d) oxygen and other oxidising agents make most flammable substances easier to ignite, burn faster and become more difficult to extinguish. Many materials that are not classified as dangerous substances will become dangerous in the presence of an oxygen-enriched atmosphere or an oxidising agent. Where oxygen or other oxidising agents are used near to dangerous substances or combustible materials, the assessment must take into account the enhanced flammability properties.

ACOP 5

Risk assessment of non-routine maintenance and related higher-risk activities

111 Employers must carry out a risk assessment before undertaking any non-routine activity which potentially increases the risk, such as process scale-up, maintenance, repair, modification, extension, restructuring, demolition or cleaning:

(a) in areas where dangerous substances are present or liable to be present, including where they are used, stored or produced;
(b) on equipment that has contained a dangerous substance.

ACOP 5

112 Employers must identify and take into account:

- **(a) the types of dangerous substance that may be present or that may become dangerous as a result of the work activity;**
- **(b) the fire and explosion hazards arising from the proposed work;**
- **(c) the necessary control and mitigation measures to enable the work to be carried out safely;**
- **(d) the appropriate system of work to ensure that the control and mitigation measures essential for safety are properly understood and implemented;**
- **(e) if written instructions need to be provided for non-routine tasks to ensure appropriate control and mitigation measures are implemented. This could be a written permit-to-work system or method statement/ instructions for entry to nominally empty vessels for inspection for example.**

Guidance 5

113 In addition to 'normal' (ie routine) activities such as storage or manufacturing processes, some activities (such as less-routine or less-frequent maintenance and repair and cleaning) may require specific procedures that expose workers to risks from dangerous substances. For example, dismantling equipment containing dangerous substances or introducing ignition sources into a hazardous area. The lack of familiarity and practice alone may increase risk of errors.

114 Non-routine activities should be considered and included in a risk assessment, as they may create risks not normally present in the place where the activity occurs and may affect the area classification (see regulation 7). It may be possible to remove the dangerous substance before the non-routine work activity starts or take special control measures to prevent the release of any dangerous substance during the work. Any additional risks associated with the activity should be assessed before work starts.

115 Factors which should be considered in the assessment for any maintenance, repair, modification, extension, restructuring, demolition or cleaning activities include:

- (a) the materials that are being used or may have been used in the area or plant where the activity is to be carried out;
- (b) which materials are dangerous substances or may become hazardous under the conditions of the proposed work. This includes residues or by-products that may occur or build up inside plant or any materials that could be released by the proposed activity, for example:
  - (i) combustible liquids with a flashpoint above 60 °C but below the temperature of the hot work during which they can evaporate to give rise to an explosive atmosphere (eg heavy fuel oil or olive oil);
  - (ii) some materials may not be classified as flammable/combustible and will have either an extremely high or no flashpoint assigned, but may ignite under certain conditions;
  - (iii) combustible dusts which may be dispersed to give rise to an explosive atmosphere or cause latent smouldering hazards (eg wood dust);
  - (iv) any substance that can decompose under the conditions of the hot work to give off flammable components which may then give rise to an explosive atmosphere (eg rubbers or plastics); and
  - (v) any substance that can decompose under the conditions of the hot work to give rise to hazardous heat or pressure effects (eg dinitrotoluene or sodium hydrosulphite or residues and by-products from reactions and other similar processes);
- (c) potential heat or ignition sources that may arise during the proposed activity;
- (d) how and where explosive atmospheres can arise;

Guidance 5

(e) the consequences of a fire or explosion during the activity;
(f) the basis of safety during the proposed activity;
(g) the training and level of competence required by the operatives;
(h) what additional protective and emergency equipment is required; and
(i) what systems of work will be needed to implement the necessary control measures during the proposed activity.

Cleaning

116 When considering the risk from cleaning activities, as with any other activity, the employer should consider the type of substance that is needed and avoid the use of dangerous substances so far as reasonably practicable.

117 If a dangerous substance needs to be used for cleaning, then the risk of it generating a hazardous area and the presence of potential ignition sources needs to be considered.

118 Where dangerous substances are introduced into plant or equipment for cleaning purposes employers should assess any additional hazards, including considering their compatibility/reactivity with other dangerous substances present.

ACOP 5

The effect of measures already in place or which will be taken as a result of DSEAR

119 Assess how the current measures, along with those to be taken, will impact on the risk and safeguard employees and others who may be affected by an incident involving the dangerous substance. In carrying out this assessment, the employer should follow the hierarchy required by these Regulations and consider in order:

(a) the effect of the measures aimed at preventing an incident;
(b) the effect of the control measures aimed at preventing its escalation; and
(c) the effect of the mitigation measures to limit the effects of an incident, including the procedures to deal with accidents, incidents and emergencies.

120 When considering what measures are reasonably practicable to address hazards arising from the presence of dangerous substances, and when assessing design of plant, equipment and workplaces, all relevant factors should be taken into account including:

(a) the adequacy of separation by distance or barrier;
(b) the design standards for the installation together with those for inspection and maintenance;
(c) protection from unauthorised access;
(d) adequate distance from potential ignition sources so that any gas or vapour from any dangerous substance will have dispersed sufficiently to be rendered non-flammable before reaching these;
(e) protection of the storage or process area from fires occurring elsewhere including the spread of fires or explosions through interconnected plant and equipment and to other parts of the premises;
(f) ensuring thermal radiation effects from fires in the locality do not threaten dangerous substances;
(g) avoidance and minimisation of:
  (i) risk of spillage;
  (ii) explosive atmospheres within and outside of plant and equipment;

ACOP 5

- (iii) **unintentional or uncontrolled chemical reactions;**
- (iv) **ignitions of dangerous substances and explosive atmospheres;**

(h) **provision of safe access to the emergency services for firefighting and rescue;**

(i) **minimising the number of people exposed to any potential explosion and the risk of a fire preventing or delaying the escape of individuals;**

(j) **training and supervision of site operatives, taking into consideration incidents and emergencies.**

The likely presence and persistence of explosive atmospheres, and the need for hazardous area classification

121 As part of the risk assessment, the employer must assess whether an explosive atmosphere is likely to form and how long it is likely to remain.

Guidance 5

122 Gases, vapours, mists and dusts can give rise to explosive atmospheres when dispersed in certain concentrations in air. The risk assessment carried out under regulation 5 informs the hazardous area classification and preparation of an area classification plan (required by regulation 7) and is intended to identify places where, because of the potential for an explosive atmosphere, controls over sources of ignition are required. The results of the classification or zoning are then used to control the equipment that may be used, or the work activities that may be carried out in these areas so as to prevent ignition. When considering area classification, employers should consider the likelihood of releases of an explosive atmosphere as well as the quantity of such releases.

123 To identify hazardous and non-hazardous areas, and then subsequently to assign zones to those areas classified as hazardous, an assessment should consider matters including:

(a) the hazardous properties of the dangerous substances involved;
(b) the amount of dangerous substances involved;
(c) the work processes, and their interactions, including any cleaning, repair or maintenance activities;
(d) the temperatures and pressures at which the dangerous substances will be handled;
(e) the containment system and controls provided to prevent liquids, gases, vapours or dusts escaping into the general atmosphere of the workplace;
(f) any explosive atmosphere formed within an enclosed plant or storage vessel; and
(g) any measures provided to ensure that any explosive atmosphere does not persist for an extended time, eg ventilation.

124 When special precautions are required, eg to allow short-term maintenance or repair, there is no requirement to draw up a revised area classification plan but there is still a requirement to take a proportionate approach to risk assess the short-term conditions.

125 Some potential sources of release may be so small that it is not necessary to specify a zoned area (see regulation 7 for guidance on hazardous area classification). This will be the case if the consequence of an ignition following a release is unlikely to cause danger to people in the vicinity. For example, if a dangerous substance is being carried through a seamless pipe, and that pipe has been properly installed and maintained, it is extremely unlikely that the substance will be released – so an explosive atmosphere would not be expected to occur from this source and the area surrounding the pipe would not be considered as hazardous.

Guidance 5

126 However, in the wrong circumstances ignition of quite small quantities of flammable gas/vapour mixed with air can cause danger to anyone in the immediate vicinity. Where this is the case, as in a relatively confined location from which rapid escape would be difficult, area classification may be needed even where quite small quantities of a dangerous substance are present.

127 Likewise, dangerous substances in small pre-packaged containers for sale, display etc in retail premises would not normally require the area to be classified as hazardous. However, procedures to clean up and dispose of any spillage/release and control ignition sources in the event of such a release would be needed.

128 Additional information relating to the process rather than the substance should also be taken into account. Some substances do not form explosive atmospheres unless they are heated; some liquids if released under pressure will form a fine mist that can explode, eg hydraulic fluids in high-pressure lines as with gas turbines.

129 Taken together these factors are the starting point for hazardous area classification, and should allow for the identification of any zoned areas.

130 For further information on the use of signs to warn of areas where an explosive atmosphere may occur, please refer to regulation 7(3).

ACOP 5

Consideration of ignition sources

131 As part of the risk assessment for their work areas employers must identify which ignition sources, including electrostatic discharges, may arise with the potential to cause a fire, explosion, energetic chemical decomposition or similar event and should be controlled. The likelihood of any potential ignition source occurring should also be considered. Certain ignition sources may have to be present if required in the process, but employers must identify and consider all possible ignition sources in areas where dangerous substances are present.

132 As a minimum, the following forms of energy should be included when considering potential ignition sources but there may be others:

(a) heat;
(b) electrical;
(c) mechanical;
(d) chemical.

133 Employers should:

(a) consider all sources of ignition appropriate to their work activities;
(b) plan to introduce (under regulation 6) measures to prevent those ignition sources occurring where they could cause harm;
(c) ensure ignition sources do not come into contact with dangerous substances or explosive atmospheres. This includes preventing mobile sources of ignition (people/equipment) moving into an area where dangerous substances are present.

Guidance 5

134 Flammable substances (particularly when in the form of an explosive atmosphere) are readily ignited. An ignition source is a release of energy, often of short duration and localised, which can ignite dangerous substances in the presence of air.

Guidance 5

135 The information obtained from the assessment of the hazardous properties of the substance and provided by the supplier should be considered together with the identified sources of ignition to assess how likely it is that the particular ignition source will ignite the dangerous substance present. The reference to 'active' in this regulation means that the potential sources present are capable of causing ignition. For static electricity, this means that a static charge is present and is capable of discharging. 'Effective' means that the energy of the potential source of ignition is sufficient to ignite the particular dangerous substance present. In the case of static electricity it means that the level of charge is sufficiently high to ignite the substance present.

136 When identifying potential ignition sources, employers must take into account the properties of the dangerous substance and the manner and state in which it is, or might foreseeably, be kept and handled. On heating, some dangerous substances could create a hazard as a result of auto-ignition, self-decomposition or an exothermic reaction. Employers should consider the effect of heat from sources such as steam pipes, heaters, flames, processing etc. Employers should also consider how the ignition of combustible materials, including packaging or rubbish, could occur and the possible escalation to involve any dangerous substances.

137 Many sources of ignition are easy to identify. Examples of potential ignition sources include:

(a) heat energy, eg:
- heating installations;
- internal combustion engines;
- open fire and flame;
- hot surfaces;
- smoking;
- hot work, including welding spatter, laser or other intense radiation sources;

(b) electrical energy, eg:
- electrical lighting devices such as lamps;
- electromagnetic radiation;
- radio frequency sources;
- short circuit;
- electrical arc;
- earth fault;
- conductor fault;
- lightning strike;
- discharges of static electricity;
- loose contact;
- excessive temperature rise due to overload;
- induction heating;
- resistive heating;
- connection to inappropriate electrical supply;

(c) mechanical energy, eg:
- friction (eg overheating);
- ultrasonic;
- impact;
- grinding;
- compression (including adiabatic compression and shock waves);

(d) chemical energy (refer to SDS/technical information sheets or data), eg:
- self-heating;
- impact- and heat-sensitive materials (eg pyrophoric substances);

Guidance 5

- reactions between dissimilar metals (eg thermite reaction sparks);
- runaway exothermic reaction.

138 In considering whether controls on ignition sources are required, employers may take into account other control measures or the likely size of a fire.

ACOP 5

Scale of the anticipated effects and extent of harm

139 Employers must consider the likely scale of a fire, explosion or other event and the potential consequences. The risk assessment should be proportionate to the risks from the quantities and nature of the dangerous substances present. The contrast between localised easily limited effects and potentially large spreading damage should be reflected in the risk assessment.

Guidance 5

140 Where there is potential for an explosion, the scale of effects and the extent of harm will depend on:

(a) the substance, the amount involved and how quickly it can be consumed. Internal building configuration or obstructions will have an effect on the rate of burning;
(b) the size of the potential explosive atmosphere and the magnitude of the direct and indirect forces created;
(c) the amount of heat radiated;
(d) how the incident could escalate and whether conditions exist or could develop to cause a further fire, explosion or similar event.

141 Consideration of these factors will allow an assessment of who will be affected by an accident, and to what extent, and what mitigation measures will be required. Providing the risk assessment has shown that there is little or no risk of injury to people, controls on ignition sources and mitigation measures may not be required. In these cases the basis of safety should be detailed in the risk assessment for the activity to justify the level of precautions to be taken.

ACOP 5

The need to consider places connected via openings to places in which explosive atmospheres may occur

142 The employer should consider the possible consequences of a potentially explosive atmosphere, or the resultant effects of any explosion, spreading through interconnected plant or entering a room, building or other enclosure where the plant is located via any openings.

Guidance 5

143 The assessment should consider areas away from the source of the hazard to which an explosive atmosphere may spread, for example through ducts. Such areas should be included in the classification system for places where explosive atmospheres may occur.

ACOP 5

Additional information for the risk assessment

144 For many mature activities, additional information to inform the drafting of the risk assessment may be readily available and should be used. If an employer is planning to undertake novel activities using new emerging technologies, further research may be required into their potential fire and explosion risks together with a description of steps taken to ensure those changes to their activities are adequately managed.

Guidance 5

145 Additional information could include details of:

(a) the skills, knowledge and experience of employees and their representatives;
(b) the training and supervision of employees;
(c) activities in adjacent areas or on adjacent premises, particularly where this could present an ignition risk; and
(d) possible misuse of dangerous substances, for example, to burn waste.

146 If the dangerous substances present are also a risk to the health of employees and others (ie they are acutely toxic, carcinogenic etc), employers should also assess health risks as required by COSHH.

ACOP 5

Reviewing and revising the risk assessment

147 Employers should plan to review their risk assessment at regular intervals. The time between reviews depends on the nature of the risk and degree of change likely in activities. It should also be reviewed if significant changes have taken place or the employer concludes it is no longer valid and following an accident or dangerous occurrence.

148 When reviewing their risk assessment employers should take the opportunity to re-examine their control and mitigation measures. This should include considering whether it is now possible to replace the substance or process with a less dangerous one. The risk assessment should be modified if developments mean it is no longer valid. Records, where required, of significant findings should also be updated.

Guidance 5

149 Employers may find it useful to note the next planned review date each time the risk assessment has been reviewed.

150 When making any change to processes and equipment involving dangerous substances, the employer should assess the effects of that change on the safe operating conditions already established. For most significant changes, it will be obvious that there are consequences for safety and that a reassessment is necessary but relatively minor changes can also lead to unsafe working conditions. For example, changing the supplier of a coating material that is applied to objects and then dried in a heated oven might be overlooked as requiring a reassessment. The new raw material, while providing the same colour effect may contain a different or higher level of solvent. This could alter the amount of vapours released into the oven raising the concentration from below the LEL to within the explosive range creating a hazardous situation.

151 Changes in the workplace which should require a risk assessment to be reviewed include:

(a) changes to the substances used;
(b) replacement or modification to the plant and/or equipment used;
(c) changes in processes or methods of work which could affect the nature of hazards and risks; and
(d) changes in the workforce – such as reductions in numbers or experience of employees involved in a work activity.

152 Adverse events such as accidents, dangerous occurrences or near misses should be a trigger for reviewing the original risk assessment.

ACOP 5

Recording the significant findings of the risk assessment

153 Where an employer employs five or more people, they should record the significant findings of their risk assessment. This should help ensure all the necessary aspects of managing the risks are covered (highlighting any gaps and actions to remedy them). The amount of information recorded should be proportionate to the level of risks present in the workplace.

154 The recorded risk assessment should follow regulation 5(4) and:

(a) describe the relevant workplace activity;
(b) identify the dangerous substances in the workplace and the risks they present;
(c) identify how risks arise including heat effects on the substances and how all risks impact on those affected;
(d) record additional information where an explosive atmosphere may occur;
(e) take into account the effects of all measures, including those under DSEAR which have been or will be taken to eliminate or control risks. This includes:
  (i) zoning and hazardous area classification;
  (ii) equipment used;
  (iii) co-ordination between employers;
  (iv) verification of overall explosion safety by a competent person as required by regulation 7(4).

155 Where the DSEAR risk assessment is integrated into an overall assessment, it can be recorded as part of that assessment.

156 The risk assessment should be completed and recorded as soon as practicable after the assessment is made, and should be stored on media that is readily accessible. If the risk assessment record cannot be completed for any reason, then a precautionary approach should be taken to ensure employee safety, and a clear timescale for the completion of the assessment documented.

157 For risks which are more significant, for example on complex sites or installations, the risk assessment may summarise measures described more fully in other referenced documents. The risk assessment should adequately outline the content of the other references and these reference documents should be readily available on site.

Guidance 5

158 All employers must carry out a risk assessment, but a record of the significant findings is only required where they employ five or more people. Although employers with fewer than five employees are exempt from this requirement, they may still find it useful to record the significant findings of their assessment, including preventive measures taken to control risk in accordance with these Regulations. A chemical works would be expected to carry out, for example, detailed hazard and operability studies, whereas a small retail outlet selling a few aerosol cans would be expected to record much less information.

159 The record should provide a description of the hazards and risks from dangerous substances which lead employers to take the relevant actions to protect safety. Where appropriate it should be linked to other health and safety records or documents describing procedures and safeguards, particularly the record of risk assessment made under the Management Regulations (see Appendix 2), COSHH, and the written health and safety policy statement required by section 2(3) of the HSW Act. It may be possible to combine these documents into one health and

Guidance 5

safety management document. It should be readily retrievable for use by management in reviews, for safety representatives or other employee representatives, and for visiting inspectors. You may find it helpful to have a system of document version control in place.

160 The amount of information that should be recorded depends on the level of risk present in the workplace. In cases where a dangerous substance poses little or no risk, it may only be necessary for employers to record:

(a) the identity of the dangerous substances present and the risks they present;
(b) the measures taken under DSEAR; and
(c) if appropriate, because of the safeguards taken, an explanatory statement that a further detailed assessment is unnecessary.

161 However, where dangerous substances in the workplace present a greater risk, the assessment record should be more comprehensive. It should include:

(a) the preventive measures in place to control the risks, including those required by regulation 6 (this can include reference to measures described more fully in other documents);
(b) enough information to demonstrate that the workplace and work processes are designed, operated and maintained with due regard to safety;
(c) information showing that adequate arrangements have been made for the safe use of work equipment, in accordance with the Provision and Use of Work Equipment Regulations 1998 (SI 1998/2303) (PUWER) (see Appendix 2).

162 For the workplace the record should show, for example, that its design allows for a process to be carried out safely, such as by the provision of a storage area for dangerous substances, or to allow sufficient space to segregate incompatible substances.

163 For work equipment, the record should show that equipment is suitable for work with the dangerous substance(s) involved.

ACOP 5

Information to be recorded when an explosive atmosphere may occur

164 EPS applies to devices, controlling devices and regulating devices contributing to or required for the safe functioning of equipment or protective systems within a potentially explosive atmosphere. It applies whether or not they are intended for use inside or outside that atmosphere and such devices and equipment should be identified and their details, purpose and maintenance arrangements should be recorded.

Guidance 5

165 Equipment intended for use in an explosive atmosphere can be shown to be suitable by reference to the classification of hazardous places into zones under regulation 7(1), and evidence of the selection of a suitable category of equipment in accordance with the manufacturer's or supplier's instructions.

166 For safety and economic reasons, it will often be preferable to install such devices in a non-hazardous area where possible. Examples include:

(a) a pump, pressure regulating device, backup storage device etc ensuring sufficient pressure and flow for feeding a hydraulically actuated safety system within the potentially explosive atmosphere;
(b) a remote controller unit connected to sensors within the potentially explosive atmosphere, eg measuring temperature, pressure, flow, gas concentration

Guidance 5

etc, which is designed to provide executive actions on one or more items of equipment or protective systems within the potentially explosive atmosphere.

167 EPS does not apply to devices (including safety, controlling and regulating devices) that do not contribute to, nor are required for, the safe functioning of equipment or protective systems within a potentially explosive atmosphere. Examples of these devices include gas detection systems providing an alarm signal and initiation of action outside the hazardous area (such as operation of an emergency ventilation system) and water spray systems designed to protect plant from fire. In this case, record sufficient information to demonstrate that the design, operation and maintenance of the device will provide the safety function required.

ACOP 5

Records concerning co-ordination

168 Where there are two or more employers at a workplace where an explosive atmosphere may occur, the risk assessment must detail the clear agreement between those employers so that the aims and requirements of regulation 11 (duty of co-ordination) have been met.

Guidance 5

169 Regulation 5(4)(c)(iv) requires that the record should explain the purpose of the co-ordination measures required by regulation 11. For example, the aim could be to alert employees of another employer to the presence of hazardous substances or places, or to facilitate emergency arrangements in the event of an accident.

170 The record should also show the arrangements the employer has in place to achieve the aims. This is likely to require reference, for example, to instructions given to other employers or their employees. It will also include information for contractors when they first start on site, including a specification of the work to be done and arrangements for supervision, and handover procedures for particular items of plant or parts of the premises.

171 The duty extends only to the potential for one employer to create a flammable atmosphere which could affect another employer's staff and matters flowing from that risk. The practicalities of implementation of the arrangements are dealt with under regulation 11.

ACOP 5

Risk assessment of new work activity

172 A risk assessment must be undertaken before any new work activity involving dangerous substances begins. This includes risks that may arise from handling, storage, plant and equipment modification, treatment and disposal of dangerous waste and by-products.

Guidance 5

173 For a new work activity, the employer should record the significant findings of the risk assessment as soon as is practicable after the assessment is made. In some circumstances, further information may be needed before the significant findings can be resolved and fully recorded. Examples of such situations include:

(a) in a research/development process setting or during a crisis with dynamic changes to events and reaction;
(b) situations where product formed is not as expected or predicted;
(c) delivered reagents are not what they were supposed to be or contain impurities that affect the reaction; or
(d) in circumstances where there is a pilot operation which must be run for a period before being assessed completely.

Guidance 5

174 The employer should update the findings as soon as the information becomes available. In the meantime the employer should adopt a precautionary approach, taking additional steps to safeguard employees.

Regulation 6 Elimination or reduction of risks from dangerous substances

Summary of regulation 6

Regulation 6 sets out how to eliminate or reduce risk to people's safety from the presence of dangerous substances by removing or controlling risks, and by providing measures to limit or mitigate the consequences for people, should an incident occur.

A hierarchy of control measures is set out in regulation 6(4) and the measures required to reduce the effects of an incident are listed in regulation 6(5). The employer has a duty to provide and to maintain the measures.

The measures in Schedule 1 are an integral part of regulation 6 and for this reason the Schedule has been incorporated within this section.

Regulation 6

(1) Every employer shall ensure that risk is either eliminated or reduced so far as is reasonably practicable.

(2) In complying with his duty under paragraph (1), substitution shall by preference be undertaken, whereby the employer shall avoid, so far as is reasonably practicable, the presence or use of a dangerous substance at the workplace by replacing it with a substance or process which either eliminates or reduces the risk.

(3) Where it is not reasonably practicable to eliminate risk pursuant to paragraphs (1) and (2), the employer shall, so far as is reasonably practicable, apply measures, consistent with the risk assessment and appropriate to the nature of the activity or operation –

- *(a) to control risks, including the measures specified in paragraph (4); and*
- *(b) to mitigate the detrimental effects of a fire or explosion or the other harmful physical effects arising from dangerous substances, including the measures specified in paragraph (5).*

(4) The following measures are, in order of priority, those specified for the purposes of paragraph (3)(a) –

- *(a) the reduction of the quantity of dangerous substances to a minimum;*
- *(b) the avoidance or minimising of the release of a dangerous substance;*
- *(c) the control of the release of a dangerous substance at source;*
- *(d) the prevention of the formation of an explosive atmosphere, including the application of appropriate ventilation;*
- *(e) ensuring that any release of a dangerous substance which may give rise to risk is suitably collected, safely contained, removed to a safe place, or otherwise rendered safe, as appropriate;*
- *(f) the avoidance of –*
  - *(i) ignition sources including electrostatic discharges; and*
  - *(ii) adverse conditions which could cause dangerous substances to give rise to harmful physical effects; and*
- *(g) the segregation of incompatible dangerous substances.*

Regulation 6

(5) The following measures are those specified for the purposes of paragraph (3)(b) –

(a) the reduction to a minimum of the number of employees exposed;
(b) the avoidance of the propagation of fires or explosions;
(c) the provision of explosion pressure relief arrangements;
(d) the provision of explosion suppression equipment;
(e) the provision of plant which is constructed so as to withstand the pressure likely to be produced by an explosion; and
(f) the provision of suitable personal protective equipment.

(6) The employer shall arrange for the safe handling, storage and transport of dangerous substances and waste containing dangerous substances.

(7) The employer shall ensure that any conditions necessary pursuant to these Regulations for ensuring the elimination or reduction of risk are maintained.

(8) The employer shall, so far as is reasonably practicable, take the general safety measures specified in Schedule 1, subject to those measures being consistent with the risk assessment and appropriate to the nature of the activity or operation.

ACOP 6

Overall approach

175 Where it is necessary to work with dangerous substances, employers are not expected to eliminate all risk but to reduce risks and to implement measures to control the remaining risks and mitigate the consequences of any fire or explosion or other harmful physical event that could foreseeably arise so far as reasonably practicable.

Substitution

176 Employers should first consider eliminating the risk if a suitable non-harmful (or, failing that, a less harmful) substitute for the dangerous substance is feasible or if a safer process exists. All aspects of the properties of the proposed substitute must be considered when substituting a dangerous substance, and the risks balanced against all the overall risks, not just its flammability or explosion properties.

177 A substance that is less flammable may not be a suitable alternative if it were of higher toxicity or more harmful to the environment than the original substance.

Control and mitigation measures

178 Having considered whether risk can be eliminated, eg by substitution, the employer should next give consideration to risk control measures before finally considering mitigation measures. Regulations 6(3) and 6(4) should be considered together when selecting control measures. The measures specified in regulation 6(4) should be applied subject to reasonable practicability and the information from the risk assessment. The measures should be applied in the order of priority set out in regulation 6(4). The list of measures is not exhaustive. There might be other effective and appropriate measures.

ACOP 6

179 If the measures set out in regulation 6(4) or other measures devised by the employer do not adequately address the risk, employers should then, so far as reasonably practicable, consider the application of the mitigation measures as set out in regulation 6(5). Regulation 6(5)(f) must be a measure of last resort.

Reduce the amount of dangerous substance to a minimum

180 Only the minimum amount of dangerous substances needed for the work activity should be kept in process areas, workrooms, laboratories and similar working areas. Dangerous substances that are not in use should be returned to the designated storage area. For dangerous substances in closed containers at retail premises, the quantity stored at point of sale must be kept to a minimum consistent with the needs of the business.

Guidance 6

181 The extent of harmful effect from fires or explosions is directly related to the quantity of dangerous substance involved. Employers should ensure their work procedures, including the selection and design of plant, are consistent with minimising the quantity of dangerous substances present in process areas, workrooms, laboratories and similar working areas. Employers should especially consider the risks posed by transfer operations involving dangerous substances and preference should be given to piped and enclosed delivery and export systems to minimise the quantity of dangerous substance in these process areas etc. Where the dangerous substance is necessarily present as feedstock and/or product, this should be the minimum necessary for the production activity and, as a guide, should not exceed that required for use or produced during half a day or one shift.

182 Employers should use plant and equipment with the smallest capacity consistent with operational needs. Replacing a batch process with a semi-batch process, or process intensification would result in a reduced inventory of dangerous substances.

Storage of flammable liquids in process areas, workrooms, laboratories and similar working areas

183 Many work activities will require the convenient availability of flammable liquids and/or flammable liquid-based products. To facilitate this, a limited quantity in suitable closed vessels may be stored in suitable cabinets or bins of fire-resisting construction and which are designed to retain spills (capacity should be 110% volume of the largest vessel normally stored in it).

184 These should be located in designated well-ventilated areas that are:

(a) away from the immediate processing area where possible; and
(b) do not jeopardise the means of escape from process and other areas.

185 The flammable liquids should be stored separately from other dangerous substances that may increase the risk of fire or compromise the integrity of the container or cabinet/bin, such as energetic substances, oxidizers and corrosive materials. Sometimes these other dangerous substances may be flammable liquids in their own right or held in a flammable liquid. However, it is still inappropriate to store these in the same cabinets or bins with other flammable liquids.

186 The recommended maximum quantities that may be stored in cabinets and bins are as follows:

(a) no more than 50 litres for extremely, highly flammable and those flammable

Guidance 6

liquids with a flashpoint below the maximum ambient temperature of the workroom/working area;

(b) no more than 250 litres for other flammable liquids with a higher flashpoint of up to 60 °C.

187 These quantities are intended to be viewed as recommended maxima representing industry safe practice, rather than absolute limits. There is some flexibility, where for example the design of modern buildings and the pattern of work can make it difficult to work within these limits, eg in large or open-plan workrooms/working areas. Where the employer proposes to store quantities in excess of the recommended maxima, a robust justification should be recorded and the risk assessment should take into account:

(a) the properties of the materials to be stored or handled in the process areas, workrooms, laboratories and similar working areas (for mixed storage the worst-case situation should be applied, ie all materials in the storage cupboard or bin should be considered as being the same material as the one that has the lowest flashpoint);
(b) the size of the process area etc and the number of people working in it;
(c) the amount of flammable liquids being handled in the process area etc and the quantities of liquid that may be accidentally released or spilled;
(d) ignition sources in the process area etc and potential fire spread in the event of an ignition:
(e) exhaust ventilation provision to the process area etc and/or the storage cupboard or bin;
(f) the fire-resisting performance of the storage cupboard or bin;
(g) the arrangements for closing the cupboard or bin doors/lid in the event of a fire;
(h) means of escape from the process area etc.

188 The objective, in the event of an incident, is to ensure that people can safely escape from process and other areas. The purpose of storing dangerous substances in cupboards and bins of appropriate construction and design is to provide a physical barrier to defer their involvement in a fire. If the dangerous substances become involved, limiting the passage of fire and hot gas should allow sufficient time for safe evacuation and for the implementation of the employer's immediate emergency procedures.

189 Paragraphs 196–199, together with Appendices 4 and 5, detail the performance requirements for fire-resisting cupboards and bins. These do not specify an absolute test or standard for the cupboard or bin itself, rather they relate to nominal construction principles. Namely that:

(a) the materials used to form the sides, top, bottom, door(s) and lid are capable of providing the required fire resistance (ie 30 minutes' integrity) and reaction to fire (ie minimal risk);
(b) the joints between the sides, top and bottom of cupboards and bins should be free from openings or gaps;
(c) the lid/doors should be close-fitting against the frame of the bin/cupboard such that there is a nominal overlap between the frame and lid/doors in their closed position;
(d) the supports and fastenings should be of a material with a melting point greater than 750 °C.

190 These criteria are the minimum performance requirements for compliance with current legislation. However, there are a number of more demanding standards and design specifications, which refer to the fire performance of the complete cabinet

Guidance 6

structure, including: BS EN 14470-1 *Fire safety storage cabinets. Safety storage cabinets for flammable liquids*,[16] Factory Mutual, Underwriters Laboratories and ANSI/NFPA 30 standards.[17] Where standards go beyond the minimum requirements of UK health and safety legislation, it should be emphasised that their implementation in the UK is not a legal requirement. However, for quantities in excess of the recommended maxima, employers/dutyholders may find the use of cabinets with enhanced fire performance can help them demonstrate that the measures they are taking are sufficient to reduce the risks so far as reasonably practicable.

191 It is the responsibility of the employer/dutyholder to ensure that cabinets to any particular standard or design specification meet the minimum legal requirements. Equally, the use of cabinets with enhanced fire performance should not be seen as a substitute for providing dedicated store rooms and outdoor storage areas for the safe keeping of containers which are nominally empty or are not needed for current work.

Storage and use of compressed and liquefied flammable gas cylinders, oxygen and oxidising gas cylinders in process areas, workrooms, laboratories etc

192 In general, gas cylinders and cartridges should be kept below 50 °C as there is an increased risk of over-pressurisation and gas discharge or rupture in the event of them being subject to elevated temperatures. For example, this is a risk in the event of a fire in a building containing gas cylinders, even if the cylinders are remote from the source of the fire. The employer should therefore minimise the number of gas cylinders kept indoors. Further guidance on this is below.

193 Ordinarily gas cylinders containing dangerous substances should not be kept in process areas etc. An exception is for gas cylinders connected to portable appliances, but the number should be limited to the minimum necessary for operational requirements. Where the appliance is fixed, the gas cylinders should normally be sited in a safe location outdoors and the gas piped indoors to the appliance. Gas cylinders that are not in use (ie not connected to an appliance) should be stored in safe, secure uncongested locations in the open air that provide ready dispersal of any released gas, and prevent accumulations or entry of gas into any enclosed area. Nominally empty cylinders should also be stored in safe location outdoors to separate them from gas cylinders in use (ie connected to an appliance).

194 Exceptionally, gas cylinders may be stored indoors where there is a specific safety, security or process quality consideration. For example, for toxic or ultra-high-purity gases needed in the electronics industry, where gases have to be temperature controlled for process reasons, or where there is potential risk of deterioration/corrosion of the cylinder/cartridge, then the gas cylinders may be stored indoors. The amounts kept should be minimised and the gas cylinders should be housed/stored in a dedicated, well-ventilated, secure storeroom or cabinet/cupboard of adequate fire-resisting construction (see regulation 6(5)(b)).

195 Where the number of gas cylinders required indoors is so few that a dedicated storeroom is not justified, a dedicated cabinet/cupboard of adequate fire-resisting construction should be used. The same storeroom/cabinet/cupboard should not be used for both stored gas cylinders and those nominally in use (connected to an appliance). Nor should it be used to store other incompatible substances or materials that pose a risk to the cylinders.

196 Fire resistance is discussed at the section 'Physical barriers of fire-resisting construction – design and performance requirements' (paragraphs 266–280). Other

Guidance 6

design requirements also apply to both storerooms and cabinets/cupboards in which gas cylinders are kept, including the provision of adequate ventilation to the outside (see regulation 6(4)(d)) and the provision of appropriate measures to mitigate the effects of any potential explosion (see regulation 6(5)(c)). Where the risk assessment and/or regulation 10 of DSEAR or advice from emergency services indicates that adequate signage is required to indicate compressed gas cylinders, this may be appropriate.

197 Further advice on the storage and keeping for use of gas cylinders and cartridges should be available from the supplier or the relevant trade association, such as the British Compressed Gas Association and UKLPG. Information on the location of such cylinders should be given to attending emergency services at the earliest appropriate opportunity.

198 The employer should justify the need to house/store gas cylinders and cartridges indoors and ensure that any storeroom, cabinet or cupboard provided for the purpose meets the minimum legal requirements.

199 A number of cabinets are commercially available that meet more demanding standards and design specifications, eg BS EN 14470-2 *Fire safety storage cabinets. Safety storage cabinets for pressurised gas cylinders*,[18] and Factory Mutual, Underwriters Laboratories and ANSI/NFPA 30[17] standards. Where standards go beyond the minimum requirements of UK health and safety legislation, it is to be emphasised that their implementation in the UK is not a legal requirement, nor should the use of such cabinets be seen as a substitute for siting gas cylinders in a safe location outdoors where it is reasonably practicable to do so.

ACOP 6

Design of plant and equipment to minimise release of dangerous substances

200 Plant and equipment used to handle, store or produce dangerous substances should be designed to an appropriate domestic national or international standard (where available) to avoid or minimise any unintended release of dangerous substances. If there is no appropriate standard the employer should be able to show that the plant or equipment is fit for the purpose of containment during its expected life and during foreseeable normal and emergency conditions. Employers should ensure so far as reasonably practicable that:

(a) work processes minimise releases by use of pipework or enclosed systems and a scheme/system is in place to ensure their contents are identifiable in accordance with regulation 10;
(b) plant is corrosion and abrasion resistant, manufactured from compatible material or treated to impart resistance;
(c) loading or unloading operations and facilities are designed, located and operated to minimise the risk of leaks, spills, overfilling and the inadvertent mixing of incompatible materials.

201 Employers must also ensure that new pressure systems comply with the Pressure Equipment Regulations 1999 (SI 1999/2001) and existing pressure systems comply with the requirements of the Pressure Systems Safety Regulations 2000 (SI 2000/128) (PSSR).

Guidance 6

202 Where any plant or equipment operates at a pressure greater than 0.5 bar above atmospheric pressure there are duties on the supplier and user under PSSR. These duties are for the user to provide any person operating the system with adequate and suitable instructions for:

(a) the safe operation of the system; and
(b) the action to be taken in the event of any emergency.

203 Detailed guidance and ACOP requirements for PSSR are available in the publication *Safety of pressure systems* L122.[19]

204 Dangerous substances that give rise to a significant risk of fire during handling or processing include those classified under CHIP as explosive, oxidising, extremely flammable, highly flammable and flammable.

205 Control rooms and other occupied buildings on sites processing or handling significant quantities of dangerous substances should be positioned or designed to provide protection from potential fires, explosions and ingress of dangerous substances. Additional guidance on protecting buildings for chemical plant is contained in the Chemical Industries Association publication *Guidance for the location and design of occupied buildings on chemical manufacturing sites*.[20]

206 To minimise the risk of fire arising from the release of a dangerous substance:

(a) Ducts, trunks and casings should be designed and installed to:
  - (i) minimise condensation of vapour or deposition of solids;
  - (ii) maintain adequate velocity throughout its length with smooth inner surfaces and large-radius bends;
  - (iii) incorporate suitable inspection and cleaning access points.

(b) Plant and equipment should be designed and operated to:
  - (i) prevent unintentional accumulation of dangerous substances and their flammable residues;
  - (ii) avoid reaching a surface temperature that may cause residues to catch fire and ignite any explosive atmosphere that may be present.

207 Mitigation measures for plant and equipment processing highly flammable solids and dusts include rotary valves, explosion suppression barriers, fast-acting valves, chokes and baffles. Mitigation measures for interconnected plant and equipment processing flammable gases and vapours include flame arresters (see BS EN ISO 16852[21]), fast-acting valves and suppression barriers.

ACOP 6

208 Where plant contains openings such as inlets and outlets, these have the potential to release dangerous substance and employers must ensure:

(a) plant doors, access points or charge/discharge points are provided with interlocks, valves or systems of work to prevent or minimise release;
(b) plant is fitted with isolation valves to minimise leaks after use, to control leaks during use and to enable safe isolation of the plant for maintenance;
(c) where personnel would be exposed to danger when operating valves manually during an emergency, plant is fitted with remotely operated isolation/shut off valves (ROSOVs).

ACOP 6

Ventilation

209 Elimination or minimisation of the release of dangerous substances by using closed systems or suitable processing and handling methods should be the first consideration. Employers should ensure proportionate, appropriate measures are taken to prevent the formation of hazardous explosive atmospheres or to limit their extent. Ventilation is (and should be) designed to dilute the concentration of any dangerous substances to a safe level (below that which could form an explosive atmosphere) by providing air changes through:

(a) an adequate number of appropriately sized openings, for natural ventilation, on all external walls at high and low levels (where reasonably practicable);
(b) mechanical extract ventilation (MEV), local exhaust ventilation (LEV) and/or forced ventilation at process and storage areas where natural ventilation cannot achieve the required air change rate to safely disperse the dangerous substance(s). LEV should be provided for processes where there is unavoidable release of a dangerous substance;
(c) any required mechanical ventilation system should be monitored for continuous operation, including a flow failure detection and alarm or other system which is suitable for the plant size and configuration. For complex areas adequate air flow should be verified by flow measurements throughout each compartment.

210 Where the release of a dangerous substance could give rise to explosive atmospheres the following measures, ranked in preference order and forming a hierarchy of control, should be considered in order to dilute the concentration of foreseeable releases to a safe level:

(a) location in the open air. Where weather protection is required, it should be designed to prevent the accumulation of dangerous substances;
(b) adequate natural ventilation for any potential source of release inside any enclosure or building where the flow of air is liable to be restricted. The ventilation should be designed to dilute the concentration of foreseeable releases of dangerous substances to a safe level by maintaining the average concentration during normal operations to below that which could form an explosive atmosphere;
(c) enclosure within a cabinet or other suitable enclosure which is constructed of fire-resisting materials and directly provided with LEV exhausting to a safe place;
(d) adequate LEV, provided and positioned to prevent or minimise releases of potentially unsafe concentrations, into the work area or room;
(e) adequate mechanical general ventilation to the workspace in the event that closely positioned LEV is either not reasonably practicable or is insufficient by itself to dilute concentrations of releases of dangerous substances to a safe level.

211 The following steps should also be taken where appropriate:

(a) prevent the formation of explosive atmospheres in enclosed spaces forming part of plant, equipment or ductwork;
(b) dryers, ovens, cabinets, connecting ducts, trunks and their associated ventilation casings should be fire-resisting structures;
(c) safely disperse vaporising liquid leaks from fixed liquefied gas vessels away from vulnerable populations and locations.

Guidance 6

Ventilation: First choice – natural dispersion

212 Locating plant and storage facilities in the open air normally ensures the best possible dispersion of dangerous substances to limit the formation and extent of hazardous explosive atmospheres. Certain features may affect the ready dispersal of any releases of dangerous substances, eg buildings, pits, and structures providing weather protection. Employers should, as appropriate, ensure these features are:

(a) sufficient distance away; or
(b) of suitable design to prevent the accumulation of dangerous substances; and where necessary:
(c) the ground should be graded to direct vapours away from occupied buildings and vulnerable populations, eg to provide safe dispersal of vapourising liquid leaks from fixed liquefied gas vessels.

Ventilation: Second choice – indoors and ventilated from open air

213 Where plant and storage facilities handling dangerous substances are located indoors, the employer should ensure that ventilation is adequate to limit the formation and extent of hazardous explosive atmospheres. The greater the air flow from and to open air (natural ventilation) the better. If possible one or more solid sides to an enclosure should be removed. Ventilation should:

(a) ensure there are no stagnant or poorly ventilated areas in the building, room or enclosure containing plant or stores where the dangerous substance can accumulate to form a hazardous explosive atmosphere; and
(b) prevent the formation of such atmospheres in any other parts of the building.

214 When considering the design and size of the ventilation requirement the employer needs to consider the nature and location of potential leak scenarios – typically this will be in two parts:

(a) the ventilation required to limit the formation and extent of hazardous explosive atmospheres that might occur during normal operations, including foreseeable deviations or excursions from normal conditions; and
(b) the emergency measures required to deal with substantial leaks that might arise in the event of accident or incident (regulation 8). Regulation 6(4)(d) is specifically concerned with the ventilation requirements for normal operations and foreseeable departures from normal.

215 In determining the ventilation requirement, the employer should take account of the range of dangerous substances that may be present and the conditions under which all activities take place, including the temperature of the workplace.

216 In a workplace where combustible dusts are likely to be present, the design of the ventilation system should take into consideration any deposits that may arise. For example, dust leaks from the plant should not be allowed to build up in such quantity that, if disturbed or dispersed, they could form a hazardous explosive atmosphere.

217 Adequate ventilation can sometimes prevent the formation of an explosive atmosphere. The variations associated with work activities mean, however, that typical effectiveness is likely to be limited to a reduction in the likelihood (chances) of a hazardous explosive atmosphere forming and/or a reduction in the extent of the hazardous area (zones). There may be sufficient ventilation in some circumstances for the extent of the hazardous area to be treated as a zone 2 of

Guidance 6

negligible extent (zone 2^{NE}) because a sufficiently small volume of explosive atmosphere would have insignificant over-pressure or thermal effects if it ignited. See regulation 7 and Schedule 2 for classification.

218 Where necessary, check that the ventilation is adequate, eg congestion or obstructions may affect the air flow. Checks may be carried out using smoke or tracer gas tests. Alternatively, where a more detailed assessment is sought, computational fluid dynamics (CFD) modelling may be used if appropriate.

219 Adequate ventilation is typically taken to be that which limits the average concentration to no more than 25% of the LEL within the building, room or enclosure containing the dangerous substance.

220 When storing dangerous substances indoors, natural ventilation provided by an adequate number of appropriately sized openings on the external walls at high and low level is usually sufficient. For buildings, the openings should be provided on opposite walls to ensure through ventilation to prevent stagnant or poorly ventilated areas. Roof ventilation openings may provide the high-level ventilation. It is acceptable for compartments and rooms to have one external wall with high- and low-level ventilation, providing stagnant or poorly ventilated areas are unlikely to occur.

221 Further guidance on assessing and designing for natural ventilation is available in BS 5925,[22] *Natural ventilation in non-domestic buildings*,[23] and *Environmental design*.[24]

222 Generic advice on the ventilation requirements for flammable liquids, compressed gases and LPG is available in *The storage of flammable liquids in containers* HSG51,[13] *Guidance for the storage of gas cylinders in the workplace* GN 2,[10] and Code of Practice 7 *Storage of Full and Empty LPG Cylinders and Cartridges*.[9]

Ventilation: Third choice – mechanical ventilation

223 If sufficient natural ventilation cannot be achieved, MEV should be provided. This should be designed to ensure the space is adequately ventilated. Ventilation openings should be correctly located in the external wall(s) of the building, room or enclosure.

224 Process areas where releases of dangerous substances might be expected to occur during normal operations will typically require LEV to ensure adequate ventilation to limit the formation and extent of hazardous explosive atmospheres. Ventilation should also take account of the potential toxicity of any dangerous substance that might foreseeably be released to make sure people are not exposed to concentrations of the substance that are likely to cause injury or ill health. This requirement under COSHH will often result in a higher standard of ventilation requirement than that required to prevent or limit the extent of an explosive atmosphere. Further information on ventilation requirements under COSHH may be found in the following:

(a) *Clearing the air. A simple guide to buying and using local exhaust ventilation (LEV)* INDG408;[25]
(b) *Time to clear the air! A workers' pocket guide to local exhaust ventilation (LEV)* INDG409;[26]
(c) *Controlling airborne contaminants at work: A guide to local exhaust ventilation (LEV)* HSG258;[27]
(d) local exhaust ventilation (LEV) workplace dust and fume extraction HSE webpages at www.hse.gov.uk/lev.

Guidance 6

225 LEV (and MEV) systems should be monitored to detect any reduction in exhaust flow that would result in inadequate ventilation. Should this occur, appropriate warning and/or automatic actions should be initiated.

226 The LEV extraction openings should be located as close as possible to the sources of potential release of the dangerous substance to limit the extent of any hazardous explosive atmosphere. LEV should be provided in other locations where explosive atmospheres might accumulate. The exact position should take account of the density of the dangerous substance (which for the majority of substances will normally be at low/floor level).

227 LEV (and MEV) systems should discharge to a safe location outdoors.

Ventilation: Fourth and last choice – extraction ventilation to the entire workspace

228 Where it is not reasonably practicable, or it is considered unnecessary to provide LEV, adequate ventilation may be achieved by general MEV to the workroom. However, such ventilation systems can affect the comfort of the occupants.

Ventilation of plant and machinery

229 Plant and machinery containing dangerous substances should be provided with adequate ventilation to prevent hazardous explosive atmospheres from occurring so far as reasonably practicable. For plant and machinery purchased from within the EU, this should be confirmed by the supplier in compliance with their duties under the Machinery Directive 2006/42/EC implemented in UK law by the Supply of Machinery (Safety) Regulations 2008 (SI 2008/1597). Documentary information should show that 'machinery is designed and constructed to avoid any risk of explosion posed by the machinery itself or by gases, liquids, dusts, vapours or other substances produced or used by the machinery'.

230 Where the plant or machinery is manufactured by the employer or imported from outside the EU, the employer takes on the responsibility for ensuring its safe design and use.

231 Plant and machinery should be properly installed to ensure the ventilation operates as designed and exhausts to a safe place, a recovery unit or a disposal unit.

232 Plant burning dangerous substances (not covered by the Gas Appliances (Safety) Regulations 1995 (SI 1995/1629)) should be provided with means of purging the combustion chamber and other parts in which a hazardous explosive atmosphere might occur before starting and completing an ignition cycle.

Adequate ventilation

233 Ventilation for plant and machinery is normally considered adequate if it limits the average concentration of any dangerous substance that might potentially be present to no more than 25% of the LEL. However, an increase up to a maximum 50% LEL can be acceptable where additional safeguards are provided to prevent the formation of a hazardous explosive atmosphere. For example, gas detectors linked to emergency shutdown of the process might be used together with maintaining or increasing the exhaust ventilation on solvent evaporating ovens and gas turbine enclosures.

Guidance 6

Temporary exhaust ventilation systems

234 Temporary exhaust ventilation systems may be provided for non-routine higher-risk activities, such as cleaning, repair or maintenance in tanks and other confined spaces or in an emergency after a release. The work procedures for such activities should be carefully considered (see also paragraphs 333–343). The atmosphere should be continuously monitored to ensure that ventilation is adequate and the area remains safe. Where workers will enter the space, the ventilation should ensure that the concentration of the dangerous substance does not exceed 10% of the LEL (irrespective of the provision of suitable breathing apparatus).

235 Industry guidance on ventilation at sites storing and handling LPG has been produced by UKLPG.[9, 28–29]

ACOP 6

Containment and collection of spillages and leaks

236 To avoid contaminating other parts of the workplace in the event of a loss of containment, employers should take measures so far as reasonably practicable to:

(a) contain spillages of dangerous substances;
(b) ensure flammable vapours/gases cannot discharge from the point of spillage into other parts of the building or into other buildings on the premises;
(c) contain and collect spills or leaks likely to give rise to a hazardous situation by means of run-off to a container or to a safe place, or otherwise treat to make it safe;
(d) provide underground storage tanks for flammable liquids and associated pipework and fittings with secondary containment or leak detection to identify leaks before a hazardous situation can arise;
(e) store solids or powders in closed vessels constructed to an appropriate domestic (or international) standard where available. Granular materials contain a proportion of dust and may be stored in designated open compounds provided adequate dust control measures are in place.

Guidance 6

237 Employers should use the risk assessment carried out in accordance with regulation 5 to decide on the need, location and type of valves necessary for isolating the supply after use, during maintenance and in the event of an emergency. Additional information on the containment of large volumes of flammable liquids may be found in:

(a) *Safety and environmental standards for fuel storage sites;*[30]
(b) *Secondary and Tertiary Containment of Bulk Hazardous Liquids at COMAH Establishments*.[31]

ACOP 6

Control and avoidance of ignition sources including electrostatic discharges

238 Unnecessary ignition sources should be avoided throughout the workplace as a general principle. Ignition sources include open flames, electrostatic discharges, unprotected powered mobile plant etc. Where the risk of a flammable atmosphere cannot be eliminated, the employer may have to control risks of ignition. Consideration of ignition sources under regulation 6 should include those that employers must consider under hazardous area classification.

ACOP **6**

239 If an employer decides to create a designated smoking area for employees and visitors, this should not be sited in or near a hazardous zone.

240 Ignition sources outside the hazardous (classified) areas should also be considered where they could pose a risk of a fire or similar event spreading into a zone where there could be an explosive atmosphere.

241 In areas where the ignition of dangerous substances could affect safety, measures must be taken to avoid ignition sources occurring or being brought into those areas. The measures employers should take include:

(a) selecting and installing appropriate electrical and non-electrical equipment that has been designed to be safe in hazardous areas. All equipment for places where an explosive atmosphere may occur should meet the essential safety requirements appropriate to the equipment category as detailed in EPS;
(b) implementing inspection, testing, cleaning and maintenance regimes for equipment to minimise ignition sources occurring as a result of overheating or fault conditions;
(c) ensuring that any portable or mobile equipment brought into hazardous areas is either suitably protected or is only brought into those areas under safe conditions ensured by implementation of a permit-to-work scheme. See regulation 6(8) and part 6 of Schedule 1 of DSEAR and the ACOP text between paragraphs 306–352 (safe maintenance, repair and cleaning procedures);
(d) prohibiting the use of open flames;
(e) implementing controls and procedures to prevent the occurrence of hazardous electrostatic discharges;
(f) ensuring heating equipment installed in areas where dangerous substances are stored or used cannot act as an ignition source. Ensuring that heating equipment or storage conditions cannot cause dangerous substances to reach their auto-ignition temperature or, where relevant, their self-accelerating decomposition temperature (SADT) for packaged materials, or the onset temperature at which thermal decomposition occurs within bulked materials;
(g) preventing the accumulation of waste materials or deposits that are liable to spontaneously combust or are readily ignited. Such materials should be placed in a closed metal bin or removed to a safe place. Deposits should be removed in such a way that their removal does not create a risk of ignition;
(h) avoiding incompatible materials that could either react together to produce heat or flames or give rise to incendive sparks following frictional contact during impact, machining, grinding or polishing. The employer's assessment of where incompatible materials could occur should consider the dangerous substances being processed and the materials of construction of plant, equipment, process areas and tools.

242 In identifying potential ignition sources, the employer should consider the properties of the dangerous substance in relation to how it is handled or may accumulate, whether unintentionally or not. If accumulations of certain fugitive dusts might self-heat and potentially ignite then such accumulations should be prevented.

243 The following factors are among the most likely ways that electrostatic discharge events presenting a risk of ignition may be generated and therefore should be avoided:

ACOP 6

(a) personnel wearing clothing and/or footwear capable of generating a build up of static electricity;
(b) containers and equipment handling dangerous substances/flammable liquids that have not been earthed/bonded correctly or have been constructed of unsuitable materials;
(c) manual cleaning operations using unsuitable equipment and/or materials;
(d) incompatible reactive materials – this includes both tools and other plant as well as dangerous substances;
(e) humidity not maintained to a range that reduces static build-up.

244 Employers should ensure that all visitors who are given access to hazardous areas of the premises are wearing appropriate antistatic clothing and are made aware of the hazards before entering any location where a potentially explosive atmosphere may exist.

245 Where mechanical exhaust ventilation (MEV) is provided and after having eliminated or minimised any releases to reduce the hazardous area of any potentially explosive atmosphere, employers should then eliminate or establish controls on ignition sources by considering all relevant factors in relation to the MEV systems. This includes:

(a) for a potentially explosive atmosphere, selection and installation of equipment and protective systems designed to be safe for the zone of the hazardous area. Such equipment and systems supplied after June 2003 should meet the essential safety requirements detailed in the EPS;
(b) not siting electric motors of fans within ducts that may contain dangerous substances, where build-up of residues may lead to overheating;
(c) designing (protecting) and installing only fan impellers, bearings, pulleys etc suitable for such use if they are in a hazardous area/zone, for example inside ductwork;
(d) providing equipment, inspection testing, cleaning and maintenance schemes to minimise overheating or fault conditions with the potential to lead to ignition of the dangerous substance;
(e) equipotentially bonding MEV, to prevent electrostatic build-up;
(f) making provision to prevent the accumulation of flammable/combustible waste deposits and enabling safe access for safe cleaning and maintenance.

Guidance 6

246 Guidance on preventing ignition from non-electrical equipment, eg by use of sensors, is contained in BS EN 13463-6.[32]

247 Advice on the control of electrostatic hazards is contained in:

(a) BS 5958-1 *Code of Practice for control of undesirable static electricity. General considerations*.[33] Also in
(b) PD CLC/TR 50404 *Electrostatics. Code of Practice for the avoidance of hazards due to static electricity,*[34] which includes a range of measures applicable to various industrial situations such as petrochemical installations and flammable powder handling.

248 Some clothing, including footwear, contains materials that can generate electrostatic discharges during use. Such discharges can ignite certain types of explosive atmospheres, eg gases or vapours and even some types of dust that are very easily ignited. Employees working in these atmospheres should be provided with antistatic footwear if the assessment indicates an ignition risk. Providing the floor is not highly insulating, antistatic footwear alone may be sufficient to control

Guidance 6

the risk because the risk from electrostatic discharges from clothing can be reduced if the wearer is earthed by the footwear and suitable flooring such as concrete or steel grids. Antistatic footwear and flooring should be tested routinely and replaced if it is found that its antistatic properties have deteriorated.

249 Employees such as electricians who need insulated footwear for other purposes will need reminding to change their footwear before entering a hazardous area unless the area can be made safe for them. This is likely to be sufficient for places classified into zone 0, 1 or 2 as specified in Schedule 2. In a small number of cases special footwear may also be needed for zones 20, 21 and 22. Other antistatic clothing must also be provided if the assessment shows this to be necessary.

250 The employer must also ensure that any personal protective equipment provided for other purposes, such as to prevent contact with substances hazardous to health, will not create electrostatic discharges if used in an explosive atmosphere.

251 Antistatic or ordinary clothing should not be removed in places where an explosive atmosphere may occur. A safe area should be established where workers are able to remove or change clothing etc. Electrostatic risks can be created by personal items brought into a hazardous area, and employers should provide instructions for employees or visitors. Where necessary, employers should ensure that visitors have appropriate antistatic clothing and should ensure that the rules set out in management procedures are followed.

252 Fan impellers necessarily have to be in the path of the dangerous substances being extracted, but the source powering them should be effectively separated from the dangerous substance. Such electric fan motors should be sited outside exhaust ducting in a position where they can be readily inspected, cleaned and maintained to avoid ignition or build-up of residues leading to overheating.

253 Fans, impellers and couplings/casings within the ducting must be properly designed and installed to avoid ignition from static discharge, frictional sparking or rubbing. Where a fan and motor are located in a classified zone of a hazardous place the fan motor design should be ATEX certified as suitable for use in that zone. Where an integral MEV fan assembly is installed in a wall leading directly to outside, it will need to be ATEX-compliant for the zone. A relevant design standard for fans on MEV systems in relation to hazardous explosive atmospheres is BS EN 14986.[35]

254 The employer will need to decide whether ignition sources can be allowed into the work area on a temporary basis. This should be based on a consideration of the risks of a flammable/explosive atmosphere being formed during the time such an ignition source is present.

ACOP 6

Segregation of incompatible dangerous substances

255 Where mixtures of dangerous substances are incompatible, they should be separated or segregated to minimise the risks. Where limited space at premises means that it is not feasible to separate substances from other substances or hazards by distance alone, then segregation should be achieved by the provision of physical partitions of fire-resisting construction.

Guidance 6

256 Where incompatible dangerous substances are separated by distance, the risk assessment should demonstrate that the risk of propagation of fire between those substances is low. General guidance on storing packaged dangerous materials is

Guidance 6

contained in *Chemical warehousing: The storage of packaged dangerous substances*.[36]

257 Although DSEAR requires only incompatible dangerous substances to be segregated, the general duties under the HSW Act mean segregation of any incompatible substances is required. For example, a risk arises where corrosive materials are stored if a release of the corrosive material could cause the containers for dangerous substances to fail. Materials defined as dangerous substances will include those that have been classified because of hazardous properties such as accelerated burning or because they can release flammable gas products.

258 Mixing of incompatible dangerous substances may occur if:

(a) a dangerous substance is sent to the wrong tank;
(b) if the contents of a delivery vehicle are not the same as the dispatch note or are out of specification; or
(c) if a tank is used to store a new material before the residues of the previous contents are adequately cleaned out.

259 The risks may be controlled by measures such as:

(a) clear labelling of transfer lines;
(b) provision of dedicated transfer lines rather than temporary flexible hoses with complex valve manifolds;
(c) use of different types of couplings for incompatible products;
(d) simple checks on the contents of incoming road tankers;
(e) written systems of work concerning tank cleaning.

ACOP 6

Mitigation – avoiding propagation

260 Regulation 6(5) details the measures that 6(3)(b) requires to be applied to reduce the consequences of an incident. Premises layout and equipment installation can mitigate the effects of a fire or explosion in cases where substance quantity/properties would have particular influence on the scale or nature of an incident. The following list is not in priority order. Employers should ensure:

(a) adequate separation of process areas from:
  (i) other parts of the premises;
  (ii) the site boundary;
  (iii) bulk storage areas;
(b) measures are taken to prevent fire and explosion from spreading to other equipment;
(c) process areas are separated from the rest of the building by physical barriers that are fire-resisting structures;
(d) physical barriers are provided where they will help to prevent damage to containers, vessels, pipework and other equipment;
(e) nominally empty containers are removed from process and work areas back to a safe place;
(f) adequate separation of storage areas:
  (i) from site boundaries;
  (ii) from occupied buildings;
  (iii) from process areas;
  (iv) from fixed ignition sources and other features that pose a threat;
  (v) between other dangerous substances within storage areas;
(g) sufficient separation to allow people to escape from fire at a store and which prevents or delays fire spread;
(h) any store is constructed to protect it from unauthorised access and from

ACOP 6

fire occurring elsewhere including on the boundary;

(i) where rooms storing dangerous liquids or gases are inside buildings, they are either:

- **(i) a clearly identified dedicated room or building adequately separated from other buildings, workrooms or hazards; or**
- **(ii) fire-resisting structures (and again their contents clearly identifiable);**

(j) bulk compressed gas and liquefied flammable gas tanks are not sited within buildings;

(k) that where compressed flammable gases and liquefied flammable gases such as LPG are stored underground, they are in suitable underground reservoirs and caverns which are not sited beneath any building or similar structure where a dangerous accumulation of gas could occur;

(l) cupboards, bins, tanks, vessels and containers (whether nominally empty or not) which contain a dangerous substance are clearly identified and any openings in them are kept closed except when in use or operation or being maintained;

(m) dispensing or decanting are not carried out in a bulk storage area where other dangerous substances are stored.

Guidance 6

261 Where lack of space means that a physical barrier of fire-resisting construction must be provided to protect (or protect against) any feature, the design and performance requirements for the barrier depend on its particular function.

262 Structures required to serve as a physical barrier of fire-resisting construction where the dangerous substance is either extremely or highly flammable or where a substance is stored or used at a temperature above or near to its flashpoint should meet the relevant fire safety performance requirements detailed in paragraphs 266–280.

263 For outdoor storage, adequate separation can be achieved by locating the storage facility at an appropriate distance from other specific features or potential hazards. Alternatively, a physical barrier such as a fire-resisting wall or partition can be used. For products which are dusts, or contain a proportion of dusts, these issues are covered in *Safe handling of combustible dusts* HSG103.[37]

264 For most dangerous substances, advice on adequate separation is detailed in HSE or industry codes of practice and guidance. These may be specific to the dangerous substance (eg LPG) or be of a generic nature (eg flammable liquids).

265 Where such advice is not available or is inappropriate due to the quantity of dangerous substance stored or manner of its use, the necessary separation distances to achieve safety will need to be determined from first principles and by taking into account any additional fire mitigation measures such as water deluge systems or monitors.

ACOP 6

Physical barriers of fire-resisting construction – design and performance requirements

266 The periods of fire resistance required should be determined by assessment of the fire hazard, taking account of its anticipated duration and severity.

Guidance 6

267 Physical barriers of fire-resisting construction should be capable of maintaining adequate fire protection to allow sufficient time for evacuation and for emergency procedures to be implemented. Fire walls are a physical barrier of fire-resisting

Guidance 6

construction and may be part of a building or free-standing structures in the open air. This fire safety performance is specified in terms of:

(a) resistance to fire; and
(b) reaction to fire.

268 Definitions and technical specifications for these terms are given in Appendix 4 'Fire resistance' and Appendix 5 'Fire reaction'.

269 Although these fire safety performance standards are not mandatory, they may be suitable in cases where the dangerous substance present is known to be flammable.

270 Compliance with the fire test standards referred to above may be demonstrated by testing, or by building the structure using materials and construction methods that are capable of providing the required fire safety performance.

ACOP 6

271 Storerooms and workrooms required to be of fire-resisting construction (ie fire resisting) should meet the following minimum requirements:

(a) every enclosing element that acts as a fire-resisting physical barrier should provide a minimum of 30-minutes' fire resistance in respect of integrity, insulation and, where applicable, load-bearing capacity. 'Enclosing element' includes every internal wall (including any door), floor (other than a floor on the ground), ceiling and its associated floor (other than the top or ceiling of a single-storey building or of a top-floor room), and any external wall that serves as a fire wall;
(b) if the room is within a building that also contains residential accommodation, the partition between the two should provide a minimum of 60-minutes' fire resistance with no connecting doors or direct access between the two parts of the building;
(c) storerooms should not contain any glazed area in any fire-resisting physical barrier except as permitted in a door;
(d) any door in a fire-resisting physical barrier should be self-closing from any position. Such a door may have a glazed viewing panel provided that it does not exceed an area of 20% of the door. Any glazing should satisfy the integrity requirements, which can be achieved by using Georgian-wired glass or a proprietary fire-resisting glazing panel. The area of such panels should be kept to a minimum so far as reasonably practicable;
(e) the materials used in the construction of a store or workroom should as a minimum have a 'low risk' in respect of their reaction to fire (see Appendix 5). (For storerooms in which LPG cylinders are stored, only materials that have a 'minimal risk' should be used.) This limitation does not apply to doors and windows together with their associated frames and any provision made for explosion relief;
(f) openings in the internal partitions of a workroom are allowed, provided:
  (i) where they are to accommodate ducts, trunks and casings, these are of fire-resisting construction;
  (ii) where they are for any other purpose (such as to allow the movement of items on a production line), a fire/smoke damper should be installed that, together with its frame, can provide a minimum of 30-minutes' fire resistance in respect of integrity;
(g) the junction between each part of a fire-resisting physical barrier should be sufficiently bonded or fire-stopped to ensure that the fire resistance is not compromised.

ACOP 6

272 Cabinets, enclosures (including any oven used solely for the evaporation of dangerous substances from materials contained therein), cupboards and bins that are required to be of fire-resisting construction should meet the following minimum requirements:

(a) every side, top, floor, door and lid should provide a minimum of 30-minutes' fire resistance in respect of integrity. This requirement may be waived in respect of:
   (i) the glazed panel of any fume cabinet or glove box;
   (ii) any booths where a screen is provided for environmental or quality control purposes;
   (iii) where provision is made for explosion relief;
   (iv) cupboards and bins where provision is made for ventilation.

273 If there is need for a viewing panel on an oven used for the evaporation of dangerous substances, this should be provided by using Georgian-wired glass or a proprietary fire-resisting glazing panel. In all other circumstances, the fire resistance integrity requirements should be maintained for cabinets, enclosures, cupboards and bins which should:

(a) be constructed of materials which are, so far as reasonably practicable, of 'minimal risk' in respect of their reaction to fire (see Appendix 5);
(b) be supported and fastened to prevent structural collapse in case of fire for at least 30 minutes. The supports and fastenings should be of high melting point material (in excess of 750 °C).

274 Ducts, trunks and casings that are required to be of fire-resisting construction should meet the following minimum requirements:

(a) provide 30-minutes' fire resistance in respect of integrity and be constructed from materials that have a 'minimal risk' (see Appendix 4) in respect of their reaction to fire, so far as reasonably practicable, except at points where provision is made for explosion relief;
(b) be supported and fastened to prevent structural collapse in case of fire for at least 30 minutes. The supports and fastenings should be of high melting point material (in excess of 750 °C).

275 Cabinets, ovens, cupboards, bins, ducts, trunks and casings should be bonded or fire-stopped to prevent or retard the passage of flame and hot gases for a period of at least 30 minutes.

276 Fire walls, storerooms, workrooms, cabinets, ovens, cupboards, bins, ducts, trunks and casings must be sufficiently robust so that their integrity in respect of fire resistance will not be damaged by any foreseeable event. This includes wear and tear from normal operational activities such as collision damage from vehicles or forklift trucks and blast over-pressure when the risk assessment identifies an explosion as a likely event.

277 Where fire walls and fire-resisting structures provide containment for leaks of dangerous substance and/or prevent any escaping vapours from reaching an ignition source while still flammable, such barriers should not be perforated by any openings. They should be constructed to withstand contact with the dangerous substance in the form and quantity that might foreseeably occur in the event of an accident.

278 The reaction to fire of the external surface of a storeroom, workroom or fire wall should be to the standard required under the relevant building

ACOP 6

legislation or that appropriate to the activity being carried out on that (external) side of the barrier, whichever is the higher standard.

279 When any surface of a structure is liable to be coated with residues, the structure should be sufficiently durable so that removing the residues will not reduce its fire resistance or capacity to resist the spread of flames.

Guidance 6

280 The risk assessment should consider the need for suitable pressure-relieving devices and appropriate fire protection and take into account foreseeable events including:

(a) fires involving releases of dangerous substances from other plant;
(b) other fires on site or at the boundary not involving a dangerous substance;
(c) sudden warming of cryogenic substances as a result of normal and abnormal operating conditions or following changes in atmospheric conditions; and
(d) thermal expansion of enclosed or trapped liquids caused by atmospheric warming or fire.

ACOP 6

Measures to mitigate the effects of an explosion (including explosion relief, suppression and pressure-resistant plant)

281 Where there is the potential for an explosion to occur involving dangerous substances used, generated or otherwise present, employers must provide appropriate and sufficient protective systems to halt incipient explosions immediately and/or to limit the range of an explosion to minimise the risk. Protective systems may be either:

(a) plant that is constructed to withstand the pressure that may potentially result from an explosion without failure; or
(b) explosion protection measures taken to restrict the spread and effects of the explosion within both the plant and the workplace.

282 For highly toxic substances, explosion suppression or explosion-resistant plant should be used, or vented materials should be otherwise safety dealt with.

283 The employer should ensure any such mitigation measure has been designed, constructed, assembled and installed, and is maintained and operated, to minimise the risk of exposing people to the physical effects of the explosion, which include pressure, flame, projectiles and the operation of the mitigation measure.

284 Steps should be taken to prevent any measures provided to counteract propagation through interconnected plant failing in a manner that might expose people to the harmful effects of the explosion.

285 Protective systems provided for use after 30 June 2003 should comply with the requirements of regulation 7(2) and Schedule 3 of these Regulations. Protective systems supplied after this date should also comply with EPS; ie the supplier has the duty to ensure that the equipment satisfies the relevant essential health and safety requirements and that the appropriate conformity assessment procedure has been carried out.

Guidance 6

286 One or more of the relevant essential health and safety requirements for a protective system may be covered by a harmonized standard adopted by the European Committee for Standardisation or the European Committee for Electrotechnical Standardisation.[38–46]

Guidance 6

287 Where protective systems are supplied as an integral part of plant or equipment (eg storage vessel, bucket elevator) their conformity is assessed during the conformity assessment of the equipment they are integrated with. This should include, where relevant, compliance with the technical standard(s) applicable to the protective system(s) integrated into the plant or equipment (EC ATEX Guidelines (Third edition) Section 3.8,[47] the EPS Regulations and BS 5908-2[48]).

288 Explosion protection measures for plant and equipment processing dangerous substances include explosion relief venting, explosion suppression equipment, pressure-shock-resistant plant and pressure-resistant plant. The design of the protection measure should mitigate possible explosions by:

(a) relieving the explosion pressures and/or hot gases to a safe place outside of the workroom;
(b) suppressing the explosion before dangerous pressures build up; and
(c) safely containing the explosion without the plant rupturing.

289 Plant and equipment normally requiring explosion protection or emergency relief venting, include:

(a) ovens and dryers normally operating with concentrations of dangerous substances below 25% of the LEL but without sufficient process control to prevent deviations above 25% LEL;
(b) ovens and dryers operating with concentrations of dangerous substances above 25% of the LEL;
(c) reactors where there is a significant residual risk from the process of a runaway reaction (unless alternative protection measures are provided such as crash cooling, reaction inhibition and quenching);
(d) cyclones, dust filters and other dust-handling plant where there is a risk of ignition;
(e) spray dryers producing combustible dusts;
(f) silos storing combustible dusts;
(g) aerosol filling rooms using liquefied flammable gases;
(h) other plant and equipment in which explosive atmospheres may occur and ignition sources cannot be eliminated;
(i) storerooms for highly flammable and extremely flammable liquids; and storage of compressed and liquefied flammable gas in buildings.

290 An explosion in a classified area may affect the safety of people in an unclassified area, eg an explosion in an item of equipment may endanger someone standing nearby, even though there is normally no flammable material outside the equipment.

ACOP 6

Transport of dangerous substances on site

291 An employer should assess, plan and implement arrangements for movement of dangerous substances on any work premises which are under their control.

Guidance 6

292 This regulation applies to on-site operations only, not to public roads. Such premises would not normally include the public highway unless it forms an integral part of the work premises when special measures may be required. The Carriage of Dangerous Goods and Use of Transportable Pressure Equipment Regulations 2009 will apply off site on the public highway.

293 Under regulation 5(2)(c)(iv), the employer should have already assessed the risks associated with transport, handling and storage of dangerous substances.

ACOP 6

Waste containing dangerous substances

294 From the risk assessment under regulation 5(2)(c)(iv) employers should decide upon and then implement appropriate control measures to ensure the safety of employees and others during the handling and storage and before or during disposal of waste materials which contain dangerous substances. Where appropriate the management arrangements should include:

(a) consideration of the hazardous properties of the waste materials;
(b) prevention of waste materials from different sources and of different composition being mixed without appropriate consideration of their compatibility;
(c) safe storage in suitable containers, labelled according to their hazardous properties;
(d) procedures and precautions for the safe collection and mixing of waste materials;
(e) prompt removal from workrooms of empty and nominally empty containers which may still contain residues of dangerous substances. These should then be stored according to the requirements of full containers while awaiting disposal in an appropriate manner.

Guidance 6

295 Employers should also be aware of, and take account of, other legislation covering the disposal of waste. Guidance on waste handling and disposal can be found on the following websites:

- the Environment Agency in England www.environment-agency.gov.uk/business/sectors/wastemanagement.aspx
- the Scottish Environment Protection Agency www.sepa.org.uk/waste.aspx
- Natural Resources Wales Cyfoeth Naturiol Cymru http://naturalresourceswales.gov.uk
- Industry guidance is also available. For example, for gases, British Compressed Gases Association leaflet GN 2.[10]

Security

296 The level of security for the storage area will depend on the potential consequences of a fire and the general security already provided for the premises. Security measures will need to take into account the possibility of arson and vandalism as well as the general site fire precautions and control measures. Examples of security measures include locks on storeroom doors, welded mesh or chain-link fencing, intruder alarms, security patrols and lockable covers to filling and discharge connections.

ACOP 6

Review of control and mitigation measures

297 Existing control and mitigation measures may need to be improved, extended or replaced, using a system of experience reviews. Measures selected should be appropriate to the work activity, consistent with the risk assessment and sufficient to reduce the risk so far as reasonably practicable.

Residual risk

298 If a residual risk remains after implementing all reasonably practicable precautions in the plant, process control, mitigation and safe systems of work, workers should – as a last resort – be provided with suitable personal and respiratory protective equipment (PPE/RPE). PPE and RPE must never be viewed as a first line of defence.

Guidance 6

299 The employer, having carried out all the required assessment and subsequent implementation work, should continue to actively ensure that the conditions they have created are maintained both in terms of hardware (plant etc) and software, and in terms of people and systems such as management procedures. For example, they should ensure their standard operating procedures are followed.

300 Employers should periodically check and review the measures they have in place (as required by regulation 5(3)) to ensure that the equipment provided is maintained.

Schedule 1 General safety measures

Summary of Schedule 1 to regulation 6(8)

The measures in Schedule 1 are an integral part of regulation 6 and for this reason, the Schedule has been incorporated within this section.

Schedule 1 to regulation 6(8)

Regulation 6(8)

1 The following measures are those specified for the purposes of regulation 6(8).

Workplace and work processes

2 Ensuring that the workplace is designed, constructed and maintained so as to reduce risk.

3 Designing, constructing, assembling, installing, providing and using suitable work processes so as to reduce risk.

4 Maintaining work processes in an efficient state, in efficient working order and in good repair.

5 Ensuring that equipment and protective systems meet the following requirements –

(a) where power failure can give rise to the spread of additional risk, equipment and protective systems must be able to be maintained in a safe state of operation independently of the rest of the plant in the event of power failure;
(b) means for manual override must be possible, operated by employees competent to do so, for shutting down equipment and protective systems incorporated within automatic processes which deviate from the intended operating conditions, provided that the provision or use of such means does not compromise safety;
(c) on operation of emergency shutdown, accumulated energy must be dissipated as quickly and as safely as possible or isolated so that it no longer constitutes a hazard; and
(d) necessary measures must be taken to prevent confusion between connecting devices.

Organisational measures

6 The application of appropriate systems of work including –

(a) the issuing of written instructions for the carrying out of the work; and
(b) a system of permits to work with such permits being issued by a person with responsibility for this function prior to the commencement of the work concerned,

where the work is carried out in hazardous places or involves hazardous activities.

ACOP regulation 6(8) and Schedule 1

Design considerations

301 The workplace, including the location of equipment, should be designed, constructed and maintained to prevent releases of dangerous substances accumulating in sufficient quantity that ignition could result in a fire and/or explosion or 'other events' that may lead to injury. In particular:

(a) except where specially designed for the purpose, flammable liquid spills should be prevented from accumulating in the workplace;
(b) workplace ventilation should be sufficient to prevent accumulations of flammable vapours forming an explosive atmosphere; and
(c) combustible dusts should be prevented from accumulating to such an extent that, if dispersed and they became airborne, an explosive atmosphere would result.

302 The employer should ensure the plant and equipment and storage conditions, including arrangements for heating the workplace, will not cause the dangerous substances to ignite, self-heat or thermally decompose.

Guidance regulation 6(8) and Schedule 1

303 Where it is intended to carry out chemical reactions it will be necessary to carry out a chemical reaction hazard assessment as part of the assessment required by regulation 5. This should provide sufficient information on the reaction kinetics, quantities and rates of heat and any gas generation to allow the plant to be designed safely with the correct process controls and operating conditions. This should include, as necessary:

(a) materials of construction;
(b) vessel design pressure;
(c) agitator configuration and speed;
(d) reactant feed controls;
(e) safe operating temperatures and pressures;
(f) heat transfer and cooling rates;
(g) process instrumentation and interlocks;
(h) process venting;
(i) safe method of reaction quenching.

304 Both normal operation and the effects of foreseeable process faults should be considered during the chemical reaction hazard assessment. Further information on the assessment and control of chemical reaction processes is given in *Designing and operating safe chemical reaction processes*.[49]

ACOP regulation 6(8) and Schedule 1

Deliberate combustion of dangerous substances

305 Where plant is designed for the deliberate combustion of dangerous substances, it is necessary to prevent or reduce the risk of an explosion by implementation of appropriate control and mitigation measures. As a minimum, controls are needed to ensure:

(a) safe fuel levels before the application of an ignition source;
(b) adequate supply of air for safe combustion of fuel and/or cooling of plant;
(c) controlled fuel flow to the combustion process ensuring unburnt fuel does not accumulate;
(d) unintentional interruption of the process is corrected;
(e) plant shutdown does not result in the formation of an explosive atmosphere;
(f) plant used for disposal by combustion is suitable for purpose and is operated by adequately trained staff.

ACOP regulation 6(8) and Schedule 1

Safe systems of work

306 Maintenance, repair, modification, extension, restructure, demolition or cleaning activities should be carried out in accordance with the appropriate safe system of work, identified in the risk assessment. Before these activities take place, where it is not reasonably practicable to eliminate stocks, spillages or contamination with dangerous substances, employers should:

(a) minimise the presence of and avoid the release of dangerous substances;
(b) prevent an explosive atmosphere forming by inerting or adequate ventilation;
(c) prevent ignition sources from being introduced into the work area; and
(d) provide appropriate emergency arrangements and equipment.

Guidance regulation 6(8) and Schedule 1

Dangerous substances used as cleaning agents

307 The use of dangerous substances for cleaning purposes should be avoided wherever possible. Where this is necessary the employer should ensure that the substance with the least hazardous properties is selected.

308 For manual cleaning operations use of a dangerous substance should be minimised by applying it to an article or surface in small sections at a time. There must be adequate ventilation of the work area and so far as reasonably practicable, elimination of ignition sources. Properly designed safety containers should be used to handle and dispense dangerous substances.

309 Where dangerous substances are introduced into plant or equipment for cleaning employers should ensure that any additional hazards, including their compatibility with other dangerous substances present, are identified and appropriate control measures are implemented.

310 Cleaning plant and equipment is a hazardous activity and the employer will also, where necessary, need to create systems to:

(a) isolate plant and equipment from sources of dangerous substances;
(b) control ignition sources in any additional hazardous zones created by the work;
(c) establish acceptable concentrations of dangerous substances for particular work activities;
(d) monitor the concentration of dangerous substances within the plant and in the surrounding area;
(e) maintain concentrations of dangerous substances below predetermined safe limits by ventilation or inerting techniques;
(f) establish action limits and procedures should the predetermined limits be exceeded during cleaning work; and
(g) ensure that the plant or equipment is inspected by a competent person and is declared clean and safe for the intended work.

311 Where entry into tanks or plant is required the employer will also need to take into account the requirements of the Confined Spaces Regulations 1997 (SI 1997/1713).

312 Industry guidance on tank cleaning is published by the Energy Institute in the *Tank cleaning safety code*.[50] Inerting means rendering the dangerous substance inactive, which usually means removing air from the tank.

Guidance regulation 6(8) and Schedule 1

Systems of work

313 Employers should ensure that there is a system of work that ensures that the control measures for a particular activity are properly understood and implemented and that an appropriate level of control is in place. The level of control will depend on the risks associated with the activity and may be based on simple operating procedures, safety method statements or a permit-to-work system.

Operating procedures (low-risk activities)

314 For low-risk activities adequate control measures should be implemented through supervision or a system of work that may include the use of written operating procedures.

315 Low-risk activities are those activities that do not increase the level of risk associated with the work normally carried out in that area. They do not, for example, introduce ignition sources into the work area or create a risk of releasing dangerous materials. They may include:

(a) routine cleaning operations;
(b) dealing with small leaks and spills during normal manufacturing or handling operations; and
(c) routine machine and equipment adjustments.

Safety method statements (medium-risk activities)

316 For medium-risk activities the employer should ensure that appropriate control measures are implemented through the use of safety method statements.

317 Medium-risk activities include maintenance, repair and servicing activities carried out by employees and contractors within or near to hazardous areas or on plant or equipment containing a dangerous substance. They may involve work that releases small quantities of dangerous substances but they should not have the potential to release a significant quantity. A 'significant quantity' is considered to be one that could create explosive atmospheres beyond the hazardous areas already designated for the installation or one that could affect the health and safety of others on or off the site. Medium-risk activities are also those which do not introduce ignition sources into hazardous areas.

318 Such activities may include:

(a) leak testing tanks and lines; or
(b) hot work in areas where there are only small quantities of dangerous substances present that do not give rise to hazardous places, eg laboratories or motor vehicle workshops.

319 A safety method statement is a written procedure to cover a particular non-routine task. As well as specifying the work to be done it will also identify the hazards associated with the work and the measures necessary to control those hazards. For repetitive tasks a generic safety method statement can be used and, where necessary, modified to take into account job-specific requirements or deviations. Safety method statements are inappropriate for high-risk activities which should be subject to a permit-to-work system (see paragraphs 321–327). However, safety method statements may be incorporated into the permit-to-work system.

320 The safety method statement, whether it is prepared in-house or by outside contractors, should be clear, concise and contain the following information:

Guidance regulation 6(8) and Schedule 1

(a) a description of the task and where it is to be carried out;
(b) the sequence and method of work;
(c) the hazards identified during the risk assessment;
(d) the skills required to deal with the hazards;
(e) the precautions necessary to control the hazards;
(f) references to specific safety procedures covering known hazards;
(g) details of any isolations and any related control procedures;
(h) details of tools and equipment to be used;
(i) method of disposal of waste and debris; and
(j) details of the state or condition in which the plant or equipment will be left at the end of the activity.

ACOP regulation 6(8) and Schedule 1

Permit-to-work systems (high-risk activities)

321 Where the proposed work is identified as a high-risk activity, employers should ensure that strict controls are in place and that the work is only carried out against previously agreed safety procedures. This should include implementing a permit-to-work system issued by a responsible person. They should be sufficiently knowledgeable about permit systems and the materials, processes, plant and equipment associated with the proposed work to be able to identify all the potential hazards and precautions.

Guidance regulation 6(8) and Schedule 1

322 High-risk activities are those where the foreseeable consequences of an error or an omission could result in immediate and serious injuries, eg an explosion or a fire that immediately affects people or traps them. They will normally include:

(a) hot work on or in any plant and equipment (including containers and pipes, eg storage tank, drum, cylinder, silo, pipeline, fuel tank etc) remaining *in situ* that contains or may have contained a dangerous substance;
(b) carrying out hot work or introducing ignition sources in areas that are normally designated as hazardous due to the presence of an explosive atmosphere. (This includes places classified as hazardous under regulation 7(1));
(c) hot work in the vicinity of plant or equipment containing a dangerous substance where a potential outbreak of fire caused by the work might spread to threaten that plant and equipment;
(d) entry into, and work in, a confined space which contains or has contained a dangerous substance or where the work activity introduces a dangerous substance into the confined space; and
(e) opening or breaking into plant and equipment, or disconnecting a fixed joint that contains or has contained a dangerous substance (excluding routine activities such as charging, discharging and sampling which are themselves covered by other standard operating procedures).

323 Guidance on permit-to-work systems may be found on HSE's website at www.hse.gov.uk/safemaintenance/permits.htm and in *Guidance on permit-to-work systems* HSG250.[51]

324 In the context of DSEAR, a permit-to-work is a documented system that authorises certain people to carry out specific work within a specified time frame. It sets out the precautions required to complete the work safely and should be based on a risk assessment. It will describe what work will be done and how it will be done – the latter can be detailed in an attached safety method statement (see paragraph 314).

325 The permit-to-work requires declarations from the person authorising the work and from the person carrying out the work. Where necessary it will also require a declaration from those involved in shift handover procedures or extensions to the

Guidance regulation 6(8) and Schedule 1

work. Finally, where plant is to be put back into service, it will require a declaration from the originator of the permit that the work is complete and that the plant is ready for normal use.

326 The permit-to-work should be clearly laid out and avoid statements which could be misleading and ambiguous. It should be designed to allow for use in unusual circumstances and detail procedures if the work needs to be suspended for any reason.

327 As well as detailing the precautions that need to be taken to prevent a fire or explosion, the permit-to-work should cover the precautions that are required to control health hazards and where necessary the hazards arising from entry into confined spaces, electric shock, high-pressure systems and contact with moving equipment.

ACOP regulation 6(8) and Schedule 1

Hot work

328 Hot work and maintenance processes that involve the application of heat or generation of sparks should be eliminated wherever reasonably practicable. Where it is not possible to do so, before work commences employers should:

(a) risk assess and implement appropriate safety procedures for all activities;
(b) make plant and equipment safe to eliminate residual dangerous substances by isolation and by adequate cleaning and gas-freeing;
(c) ensure that where inerting with nitrogen, carbon dioxide or combustion gas is used, risks from inerting gas are considered under COSHH; and
  (i) inerting material is maintained at adequate levels for the duration of the work to ensure the atmosphere in the plant or equipment cannot support combustion or that any free volume is sufficiently small that any explosion within this will not pose a danger;
  (ii) a calibrated oxygen detection meter is used to ensure the oxygen concentration is adequately low and does not rise above the determined safe level;
(d) ensure a competent person inspects and monitors the atmosphere inside plant and equipment.

329 In exceptional circumstances hot work can be carried out on operationally active or inactive plant or equipment that has previously contained a dangerous substance without cleaning or inerting. Such techniques are only applicable to plant or equipment containing liquids or gases and are not suitable for plant containing dangerous substances which are solids, dusts or explosives or that contain liquid or gaseous oxygen.

330 Where it is intended to carry out hot work on plant or equipment that still contains a dangerous substance the employer must ensure that:

(a) there is sufficient liquid or gas within the plant to prevent air or oxygen from entering and forming an explosive atmosphere;
(b) flames or heat will only be applied to the outside surface of the plant;
(c) the plant cannot fail or leak as a result of the hot work activity and allow liquid or gas to escape and ignite;
(d) the gas or liquid composition cannot change to become an explosive atmosphere during the hot work;
(e) sufficient control can be exercised over the movement of materials into or out of that plant and any associated plant or equipment;

ACOP regulation 6(8) and Schedule 1

(f) substances or residues present in the plant cannot undergo any reaction or decomposition leading to a dangerous increase in pressure or attack on the metal;
(g) these techniques are only carried out under a strict permit-to-work system;
(h) all personnel involved in planning and carrying out the work and supervising it are competent and trained in appropriate procedures and fire and explosion hazards; and
(i) there are no explosive atmospheres around the work area arising from that plant or other work activities.

Guidance regulation 6(8) and Schedule 1

331 The specified conditions above should prevent a fire or explosion by ensuring that the contents of the plant are kept above their higher explosion limit and that the hot work is only carried out on the outside of the plant.

332 Eliminating dangerous substances before performing maintenance will include removing stocks of dangerous substances, cleaning and making plant safe, sealing drums and containers, isolating pipework or material handling systems and clearing up any spills or deposits of dangerous substances.

Preparation and procedures for hot work

333 Wherever reasonably practicable, employers should eliminate the need for hot work by the use of other processes that do not involve the application of heat or the generation of heat or sparks.

334 The use of cold-cutting equipment (including low-speed drills, saws and chisels) may not be considered to be 'hot work' but they may still create sparks or hot surfaces with the potential to ignite explosive atmospheres. Their use, therefore, should be assessed and controlled as for any other potential ignition source.

335 Where it is not reasonably practicable to avoid hot work on plant or equipment that has contained a dangerous substance, regulation 6(3) requires the employer to apply appropriate measures, so far as reasonably practicable, to control the fire and explosion risks.

Cleaning and gas-freeing plant for hot work

336 Before starting work, plant and equipment which has contained a dangerous substance should be isolated, cleaned and – in the case of volatile liquid and solid dangerous substances – gas-freed and ventilated to remove dangerous substances. These are hazardous operations requiring their own assessments and appropriate safety procedures.

337 Thorough removal of all residues must be ensured. However, this may not be reasonably practicable for very large tanks, for example on ships, nor may complete inerting of the enclosed spaces prior to work. In these cases, the areas surrounding the proposed repair site should be cleaned back to an extent assessed as adequate by a competent person. All involved will need to be experienced and trained in this type of work. The competent person will need to ensure that:

(a) surfaces have been cleaned of all residues of dangerous substances;
(b) there are no significant amounts trapped or held in any voids, crevices or absorbent components of the plant;
(c) by monitoring the atmosphere within the plant or equipment, it is free from all flammable gases and vapours;
(d) the concentration of any dangerous substance is less than 1% of its LEL;

Guidance regulation 6(8) and Schedule 1

(e) flammable gases or vapours do not reoccur during the hot work activity. The need for further continuous or periodic monitoring of the atmosphere throughout the work activity should be considered.

338 Where it is not reasonably practicable to eliminate dangerous substances by adequate cleaning techniques, the employer must implement measures to control and if necessary mitigate against the fire and explosion risks arising from the hot work.

Inerting

339 Gas-freeing and inerting should only be performed by those competent to do so, with appropriate measuring equipment, systems for work and safety equipment.

340 In some cases, cleaned and emptied plant and equipment may still contain residues of dangerous substance which are difficult or impracticable to remove. Inerting may be appropriate where there is a risk that these residues could ignite or form an explosive atmosphere during hot work.

341 Inerting is only applicable to flammable, highly flammable or extremely flammable dangerous substances or to substances that can create an explosive atmosphere on heating. It is not applicable to dangerous substances which are oxidisers or chemically unstable and are able to react without the presence of atmospheric oxygen to give rise to hazardous heat or pressure effects.

342 Inerting techniques may use water, nitrogen foam, nitrogen gas, combustion gas or carbon dioxide to reduce the oxygen content in the plant to below the levels that combustion can occur. Inerting may be hazardous if insufficient inert material is added to plant and equipment to achieve and maintain a non-combustible atmosphere or if people are exposed to dangerous quantities of toxic or asphyxiating gases and vapours.

343 Further information can be found in *Safe work in confined spaces* L101.[52] Additionally, the resultant displaced dangerous substances may accumulate outside the plant and equipment giving rise to unforeseen health and safety hazards so should either be vented to a safe place or to atmosphere as appropriate.

ACOP regulation 6(8) and Schedule 1

Using gas welding and cutting equipment

344 Employers must implement measures to control the risk of fires and explosions arising from gaseous welding mixtures and cutting equipment. These measures will include:

(a) providing appropriate equipment designed and constructed to recognised standards, which has been inspected and maintained in accordance with the manufacturer's instructions;
(b) protecting welding/cutting equipment, pipework and any associated fuel gas or oxygen compressed gas cylinder by the use of a suitable device which will arrest the progression of a flame flashback or acetylene decomposition;
(c) where appropriate, monitoring or detecting leaks or the possible build-up of oxygen or fuel gases in confined spaces;
(d) ensuring work takes place away from heat sources and there is adequate ventilation. If the use of gas cylinders in confined spaces cannot be avoided, supply valves should always be securely closed if cylinders are left unattended and special precautions, such as local exhaust ventilation, need to be taken;

ACOP regulation 6(8) and Schedule 1

(e) routing hoses or pipes through areas where they are not easily damaged and not near to heat sources;

(f) where moveable gas hoses or pipes are used or routed through confined spaces, they should be removed to a well-ventilated area at the end of each operation. Where this is not possible, they should be disconnected from source at a point outside the confined space and their contents safely vented; and

(g) appropriate training, instruction and supervision to ensure correct operating procedures are followed.

Guidance regulation 6(8) and Schedule 1

345 Industry guidance on storage and use of gases can be found from suppliers and from the British Compressed Gas Association.[10, 53–56] Guidance on acetylene and on the fire and explosion risks associated with hot work is available on the HSE website at www.hse.gov.uk/fireandexplosion/acetylene.htm, and in UKLPG Code of Practice 7 *Storage of Full and Empty LPG Cylinders and Cartridges*.[9]

ACOP regulation 6(8) and Schedule 1

Redundant plant and equipment

346 Before any decommissioning or relocation of fixed or bulk storage, advice should be sought from the supplier of the dangerous substance about making plant safe before it is mothballed, dismantled, transferred to a holding area or removed from site.

347 Where tanks have been made temporarily safe to be taken off site for cleaning and disposal, they should be maintained in a safe condition before and during transport and subsequent demolition.

348 Portable gas cylinders (transportable pressure receptacles) of any kind for which there is no further use should be returned to the supplier who is normally also the owner for refill or disposal. Employers should keep track of cylinders, drums and other transportable containers on site so that they may be safely disposed of.

Guidance regulation 6(8) and Schedule 1

349 The plant and equipment should first be adequately isolated from sources of dangerous substances (eg by permanent disconnection or by use of blanking plates) and drained or cleaned of residual material. Additional cleaning or inerting may be required depending on the risk assessment and the proposed method of disposal. Normally the plant and equipment should be effectively cleaned of all residues and, where necessary, gas freed before being mothballed, dismantled, transferred to a holding area or removed from site. **For any decommissioning or relocation work involving fixed bulk storage (such as compressed and liquefied flammable gas and underground petrol tanks), documentary evidence that industry guidance was consulted and advice was obtained from the supplier on safe systems should be preserved, together with any permits-to-work.**

350 Where plant or equipment containing residual product is to be removed from site without cleaning or gas freeing, the employer should ensure that it can be handled and transported safely and that those receiving it are aware of the hazards and are competent to deal with them. The employer should ensure that the risk assessment identifies the hazards and the necessary control measures for the transfer/transport procedures and that, where available, agreed industry standards are followed.

Guidance regulation 6(8) and Schedule 1

Disposal of static vessels containing compressed and liquefied flammable gases

351 Guidance on removing redundant bulk LPG vessels from sites is contained in the UKLPG Code of Practice 26 *Uplifting of Static LPG Vessels from Site and their Carriage to and from Site by Road*.[57] The minimum practicable amount of LPG that can remain in a vessel of less than five cubic metres when it is removed from site is 50 kg.

Disposal of underground petrol storage tanks

352 Where it is intended to leave the tank on site, it should be made permanently safe, eg by filling. Suitable solid materials for filling underground storage tanks *in situ* include sand/cement slurry, foamed concrete and urea amino plastic foam. Before adding the solid material the tank should be emptied of residual product and then made safe by filling with an inert material such as nitrogen foam, nitrogen gas, water, carbon dioxide or locally generated combustion gas. Alternatively the tank can be made safe before filling by suitable cleaning and degassing methods. Guidance is available from the Association for Petroleum and Explosives Administration's (APEA), *Guidance for Design, Construction, Modification, Maintenance and Decommissioning of Filling Stations* (The Blue Book) and is available from the APEA website.[58]

Regulation 7 Places where explosive atmospheres may occur

Summary of regulation 7

Regulation 7 contains specific requirements to be applied where an explosive atmosphere may occur (in addition to the requirements in regulations 5 and 6).

The regulation requires competent identification of hazardous and non-hazardous zones before new work starts and verification by a competent person.

Regulation 7

(1) Every employer shall classify places at the workplace where an explosive atmosphere may occur into hazardous or non-hazardous places in accordance with paragraph 1 of Schedule 2 and shall classify those places so classified as hazardous into zones in accordance with paragraph 2 of that Schedule; and that Schedule shall have effect subject to the notes at the end of that Schedule.

(2) The employer shall ensure that the requirements specified in Schedule 3 are applied to equipment and protective systems in the places classified as hazardous pursuant to paragraph (1).

(3) Where necessary, places classified as hazardous pursuant to paragraph (1) shall be marked by the employer with signs at their points of entry in accordance with Schedule 4.

(4) Before a workplace containing places classified as hazardous pursuant to paragraph (1) is used for the first time, the employer shall ensure that its overall explosion safety is verified by a person who is competent in the field of explosion protection as a result of his experience or any professional training or both.

(5) The employer shall ensure that appropriate work clothing which does not give rise to electrostatic discharges is provided for use in places classified as hazardous pursuant to paragraph (1).

Regulation 7

(6) This regulation is subject to the transitional provisions in regulation 17(1) to (3).

Schedule 2 Classification of places where explosive atmospheres may occur

Schedule 2 to regulation 7(1)

Regulation 7(1)

(which substantially reproduces the provisions of Annex I of Council Directive 99/92/EC)

1 Places where explosive atmospheres may occur

A place in which an explosive atmosphere may occur in such quantities as to require special precautions to protect the health and safety of the workers concerned is deemed to be hazardous within the meaning of these Regulations.

A place in which an explosive atmosphere is not expected to occur in such quantities as to require special precautions is deemed to be non-hazardous within the meaning of these Regulations.

2 Classification of hazardous places

Hazardous places are classified in terms of zones on the basis of the frequency and duration of the occurrence of an explosive atmosphere.

Zone 0

A place in which an explosive atmosphere consisting of a mixture with air of dangerous substances in the form of gas, vapour or mist is present continuously or for long periods or frequently.

Zone 1

A place in which an explosive atmosphere consisting of a mixture with air of dangerous substances in the form of gas, vapour or mist is likely to occur in normal operation occasionally.

Zone 2

A place in which an explosive atmosphere consisting of a mixture with air of dangerous substances in the form of gas, vapour or mist is not likely to occur in normal operation but, if it does occur, will persist for a short period only.

Zone 20

A place in which an explosive atmosphere in the form of a cloud of combustible dust in air is present continuously, or for long periods or frequently.

Zone 21

A place in which an explosive atmosphere in the form of a cloud of combustible dust in air is likely to occur in normal operation occasionally.

Schedule 2 to regulation 7(1)

Zone 22

A place in which an explosive atmosphere in the form of a cloud of combustible dust in air is not likely to occur in normal operation but, if it does occur, will persist for a short period only.

1 Layers, deposits and heaps of combustible dust must be considered as any other source which can form an explosive atmosphere.

2 "Normal operation" means the situation when installations are used within their design parameters.

Schedule 3 Criteria for the selection of equipment and protective systems

Schedule 3 to regulation 7(2)

Regulation 7(2)

(1) Equipment and protective systems for all places in which explosive atmospheres may occur must be selected on the basis of the requirements set out in the Equipment and Protective Systems Intended for Use in Potentially Explosive Atmospheres Regulations 1996 unless the risk assessment finds otherwise.

(2) In particular, the following categories of equipment must be used in the zones indicated, provided they are suitable for gases, vapours, mists, dusts or mists and dusts, as appropriate:

- *in zone 0 or zone 20, category 1 equipment,*
- *in zone 1 or zone 21, category 1 or 2 equipment,*
- *in zone 2 or zone 22, category 1, 2 or 3 equipment.*

(3) For the purposes of this Schedule and regulations 7(2) and 17(1) –

(a) "equipment" means machines, apparatus, fixed or mobile devices, control components and instrumentation thereof and detection or prevention systems which, separately or jointly, are intended for the generation, transfer, storage, measurement, control and conversion of energy and the processing of material, as the case may be, and which are capable of causing an explosion through their own potential sources of ignition;

(b) "protective systems" means devices other than components of equipment which are intended to halt incipient explosions immediately or limit the effective range of an explosion or both, as the case may be, and which systems are separately placed on the market for use as autonomous systems;

(c) "devices" means safety devices, controlling devices and regulating devices intended for use outside potentially explosive atmospheres but required for or contributing to the safe functioning of equipment and protective systems with respect to the risks of explosion;

(d) "component" means any item essential to the safe functioning of equipment and protective systems but with no autonomous function; and

(e) "potentially explosive atmosphere" means an atmosphere which could become explosive due to local and operational conditions.

Schedule 4 Warning sign for places where explosive atmospheres may occur

Schedule 4 to regulation 7(3)

Regulation 7(3)

(which substantially reproduces the provisions of Annex III of Council Directive 99/92/EC)

Distinctive features:

(a) triangular shape;
(b) black letters on a yellow background with black edging (the yellow part to take up at least 50% of the area of the sign).

Guidance regulation 7(3) and Schedule 4

353 The particular requirements of regulation 7 are in addition to the requirements in regulation 6 and are limited to explosive atmospheres as defined in regulation 2, ie those that may occur in air under normal ambient conditions (for the purposes of standardisation defined as -20 to 40 °C, and 0.8 to 1.1 bar). For atmospheric conditions outside these ranges, regulations 7 and 11 do not apply. However, the rest of the Regulations do apply in such circumstances and the employer is required to assess such atmospheres to ensure that the risks from fire or explosion are either eliminated or reduced so far as reasonably practicable.

ACOP regulation 7(3) and Schedule 4

Hazardous area classification

354 Where a hazardous area classification study has been carried out under regulation 7, this should be recorded in the form of a drawing which:

(a) identifies the hazardous areas and types of zones;
(b) shows the extent of the zones in both plan and elevation (ie illustrates the three-dimensional nature of the hazardous zone);
(c) is supplemented by text giving information about;
  (i) the dangerous substances that will be present;
  (ii) the work activities that have been considered;
  (iii) other assumptions made by the study;
(d) is retained as part of the documentation in support of regulation 5;
(e) is considered whenever new equipment is to be introduced into a zoned area.

Guidance regulation 7(3) and Schedule 4

355 Schedule 2 defines a place as hazardous where an explosive atmosphere may occur in such quantities as to require special precautions to protect the health and safety of workers. A place where an explosive atmosphere is not expected to occur in such quantities as to require such special precautions is deemed to be non-hazardous, but the rest of DSEAR still needs to be considered.

356 In deciding when hazardous area classification is not necessary for a small quantity of dangerous substance the actual circumstances of use and any specific

Guidance regulation 7(3) and Schedule 4

industry guidance should be taken into account. Guidance is available on HSE's website at www.hse.gov.uk/electricity/atex/classification.htm.

357 A spillage from a small bottle of solvent would release so little flammable material that no special precautions for the selection of equipment are needed and therefore the area would not be classified as hazardous. An assessment would still be required to identify the normal control measures necessary to protect the health and safety of those using the solvent. For example, measures to prevent spillage; measures to reduce exposure to fumes or vapours; controls over naked flames and similar large and continuous ignition sources when the solvent is being used. After such a spillage it would be necessary to control obvious ignition sources (eg smoking) during clean up and disposal of the liquid.

358 The expression 'special precautions' used in Schedule 2 means precautions to control potential ignition sources within a hazardous area, in particular in relation to the construction, installation and use of equipment. The term 'not expected to occur in such quantities' means that employers should consider the likelihood of occurrences of explosive atmospheres as well as the potential quantity of such dangerous substances when considering area classification. So if a release is extremely unlikely to occur and/or if the quantities released are small, it is unlikely to be necessary to classify the area as hazardous.

359 Hazardous area classification should be carried out as an integral part of the risk assessment process. Its purpose is to define the extent, frequency and duration of any occurrence of an explosive atmosphere (the zone). The zone in turn defines the requirements for the selection and installation of equipment and protective systems to prevent sources of ignition so far as reasonably practicable.

360 The hazardous area classification drawing could be in the form of two separate drawings showing a simplified approach to the configuration of the plant. Such drawings should be supplemented by text giving information about the dangerous substances that will be present, the work activities that have been considered, and other assumptions made by the study. The drawings and documents should be retained as part of the documentation in support of regulation 5. The information in these documents should be considered whenever new equipment is to be introduced into a zoned area.

361 If a maintenance process increases the risk of a release of dangerous substance, then the scope of the hazardous area may need to be enlarged. It is normally unnecessary to produce a new hazardous area classification drawing for the duration of the maintenance work. The risk assessment may find that new temporary controls and procedures are necessary for maintenance activities.

362 If during periods of maintenance, dangerous substances can be adequately and reliably excluded from an area which is normally classified, it is likely to be possible to treat the area as non-hazardous (depending on the risk assessment in the particular circumstances).

363 Trade associations can be a good source of advice on classifying areas into hazardous zones where flammable liquids and gases are present. Typical hazardous area classifications for a number of different circumstances may be found in HSE and industry publications on flammable substances.

364 Additional guidance on hazardous area classification and controlling ignition sources is contained in the following publications:

(a) Model Code of Safe Practice Part 15 *Area Classification Code for Installations Handling Flammable Fluids;*[12]

Guidance regulation 7(3) and Schedule 4

(b) *Electrostatics. Code of practice for the avoidance of hazards due to static electricity*;[34]
(c) *The storage of flammable liquids in tanks* HSG176 ;[14]
(d) *The safe use and handling of flammable liquids* HSG140;[15]
(e) Solvents Industry Association guidance and visual training aids at www.sia-uk.org.uk/health-and-safety.htm;
(f) *Safe handling of combustible dusts. Precautions against explosions* HSG103.[37]

365 An international standard, *Explosive atmospheres. Classification of areas. Explosive gas atmospheres*,[59] explains the basic principles of area classification for gases and vapours. Its counterpart for dusts is *Explosive atmospheres. Classification of areas. Combustible dust atmospheres*.[60] Taken together, these form a suitable basis for assessing the extent and type of zone, and can be used as a guide to complying with regulation 7 and Schedule 2. However, they cannot give the extent and type of zone in any particular case, as site-specific factors should always be taken into account.

366 In addition to HSE guidance, industry-specific codes containing examples have also been published by various organisations. Applied appropriately, they are valuable in encouraging consistent interpretation of the requirements. Such guidance and codes include:

(a) the Energy Institute's Model Code of Safe Practice Part 15 *Area Classification Code for Installations Handling Flammable Fluids*;[12]
(b) APEA's *Guidance for Design, Construction, Modification, Maintenance and Decommissioning of Filling Stations*;[58]
(c) BS 476 Parts 4, 6, 7 and 11;[61]
(d) UKLPG Code of Practice 1 Part 1 *Bulk LPG Storage at Fixed Installations*;[28]
(e) UKLPG Code of Practice 7 *Storage of Full and Empty LPG Cylinders and Cartridges;*[9]
(f) British Compressed Gases Association publications.[10,53–56]

367 Some repeated activities such as refuelling cars, or loading and unloading tankers intended for use on the public roads, involve the introduction of potential sources of ignition into an area where a spill is possible, which would meet the description of a hazardous area. In these circumstances, safety can be achieved by isolating power sources (eg turning off engines etc) while a transfer is taking place, and making suitable checks before and after a transfer, and before moving a vehicle into or out of a hazardous area. The risk assessment made under regulation 5 should consider the controls necessary.

ACOP regulation 7(3) and Schedule 4

Selection of equipment for use in hazardous areas

368 The employer should only use 'products' (equipment, protective systems, safety devices, components and their combinations) in potentially explosive atmospheres that comply with the specific essential health and safety requirements (EHSRs) of EPS, unless the risk assessment states otherwise.

Guidance regulation 7(3) and Schedule 4

369 In addition to the term 'hazardous area', the term 'safe area' is commonly used by suppliers of equipment that is not ATEX protected and which is used in non-hazardous areas. **Note: the suitability of a 'product' includes its potential as an ignition source by any mechanism, including heat, mechanical, chemical and electrical energy.**

Guidance regulation 7(3) and Schedule 4

370 Where the product is supplied from within the EU territory, confirmation of this should be available from the supplier/manufacturer, who in compliance with EPS should provide information and mark the product to confirm the nature of potentially explosive atmosphere it is designed for.

371 An employer may need to obtain expert advice when sourcing equipment to ensure it is suitable for the specific hazard circumstances, eg organic dusts or a chemical with particular ignition characteristics. This also applies to mobile equipment.

ACOP regulation 7(3) and Schedule 4

372 The employer should ensure that the product is installed, operated and maintained in accordance with the supplier's instructions. The employer should ensure that any modification or change of a device or component maintains compliance with the EHSRs for the product.

373 Products that were already in use within EU territory before July 2003 are not subject to the requirements of Schedule 3. These can continue in use providing the employer has assessed them and is able to demonstrate that the risks from fire or explosion are either eliminated or reduced so far as reasonably practicable. Equipment that is 'second-hand', ie it has already been used in the EU before July 2003, is not subject to the requirements of Schedule 3 but must meet the relevant requirements set out in regulation 6.

374 Where an employer intends to use the flexibility provided by the phrase in Schedule 3(1) 'unless the risk assessment finds otherwise', this decision must be adequately justified and recorded by their risk assessment which should confirm that the approach taken provides an equivalent level of safety to DSEAR. The derogation cannot be used to avoid the requirements placed on 'responsible persons' under EPS concerning the supply, importation and/or putting into use of products.

Guidance regulation 7(3) and Schedule 4

375 The effect of regulation 7, Schedule 2 and Schedule 3, taken together, is to require new equipment and protective systems provided for use at work in places classified as hazardous to comply with EPS. In most cases this can be achieved following an area classification study by selecting EPS equipment of an appropriate category according to the criteria set out in Schedule 3.

376 A standardised marking scheme is widely used to help identify equipment suitable for a specific location. Equipment built to the requirements of EPS will carry the explosion protection symbol 'Ex' in a hexagon, the equipment category number (1, 2 or 3), the letter G and/or D depending on whether it is intended for use in gas or dust atmospheres or both, and other essential safety information. In many cases this will include a temperature rating expressed as a 'T' marking, and sometimes a gas group. These indicate limitations to safe use. Employers and those installing equipment should consider the marking and documentation provided with 'Ex' equipment when it is being installed.

Guidance regulation 7(3) and Schedule 4

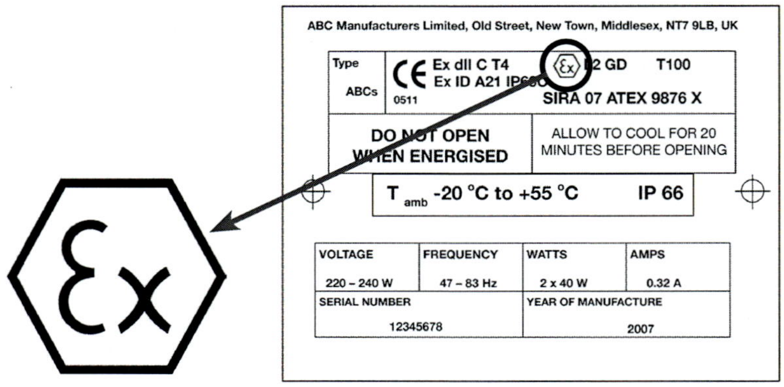

Figure 1 Typical ATEX equipment marking plate

377 In the case of equipment that is suitable for both explosive gas atmospheres and explosive dust atmospheres, the equipment should display two separate markings. For example, this would appear within the marking plate (see Figure 1) information as:

(a) II 1 G – Ex ia IIB T4; and
(b) II 1 D – Ex ia IIIC T120 °C.

378 Further guidance may be found in:

(a) BS EN 13463-1 *Non-electrical equipment for use in potentially explosive atmospheres. Basic methods and requirements*;[62]
(b) BS EN 60079-14 *Explosive atmospheres. Electrical installations design, selection and erection*;[63] and
(c) BS EN 60079-0 *Explosive atmospheres. Equipment. General requirements*.[64]

379 Equipment including flame arresters (see BS EN ISO 16852[21]), pressure valves and ROSOVs should be ATEX certified and properly installed and suitably maintained.

380 The controls apply particularly to the selection of fixed equipment that can create an ignition risk. However, the same principles may be extended to control the use of mobile equipment, other sources of ignition that may be introduced into the workplace (eg matches and lighters, car key electronic fobs, mobile phones etc) and the risks from electrostatic discharges. Advice on electrostatic discharges is contained in *Electrostatics. Code of Practice for the avoidance of hazards due to static electricity*.[34]

381 The derogation referred to in paragraph 374 is intended to allow equipment of a higher or lower category than that normally required for the zone in question to be used where:

(a) equipment is temporarily taken into a zoned area and alternative effective precautions are provided to control the risk. An example might be arrangements to isolate or shut down equipment to prevent the release of a dangerous substance;
(b) workers can be excluded from the hazardous area, and will not be at risk from any ignition of an explosive atmosphere;
(c) equipment of the required category is simply not available, but a lower category can be used in combination with other protective measures to achieve the purposes of these Regulations.

Guidance regulation 7(3) and Schedule 4

382 In addition, the derogation cannot be used to circumvent the requirements placed on 'responsible persons' under EPS, and in particular:

(a) to allow equipment imported from outside the EU, built to other standards, to be used without complying with the EPS Regulations before it is placed on the market or put into service in the European Economic Area (EEA);
(b) to justify equipment built to lower standards than that specified by EPS.

383 The leeway provided by the Schedule 3 derogation does not affect the duties placed on manufacturers, suppliers, importers and other 'responsible persons' under EPS. Users who manufacture equipment for their own use, or who import directly from outside the EEA, are considered to be a 'responsible person' under EPS and take on the full responsibility for complying with those Regulations when putting that equipment into service for the first time in the EEA.

384 Under the European Agreement concerning the International Carriage of Dangerous Goods by Road (ADR), an equipment certificate of approval – provided by the Department for Transport – must be obtained for pumps that use the vehicle's engine to power tanker offloading activities. Earthing links/clamps do not contain their own potential source of ignition and so according to the ATEX Guidelines[47] at paragraph 3.7.2 are not subject to ATEX. Nevertheless the potential for ignition must be considered including from misuse.

ACOP regulation 7(3) and Schedule 4

Marking areas containing explosive atmospheres

385 Employers should either:

(a) erect a sign complying with Schedule 4 at the points of entry to alert people to the locations where an explosive atmosphere may occur and the need to take special precautions; or
(b) if this is impractical, employers should be able to demonstrate that they have taken equally effective means to alert people to locations where an explosive atmosphere may occur and that special precautions are required for entry.

Guidance regulation 7(3) and Schedule 4

386 The requirement in regulation 7(3) applies in addition and without prejudice to similar requirements in other legislation, such as the Dangerous Substances (Notification and Marking of Sites) Regulations 1990 (SI 1990/304) and the Health and Safety (Safety Signs and Signals) Regulations 1996 (SI 1996/341). If signs have already been provided under those Regulations, and they are sufficient to warn of an explosive atmosphere, then the addition of the sign specified in Schedule 4 may not be necessary. However, if the employer considers the existing signs are not sufficient the sign in Schedule 4 may need to be applied in addition. Marking is not required for individual zones.

387 Signs are useful to identify where:

(a) special workplace or site rules apply, eg designated smoking areas, antistatic footwear to be worn, or access restricted to authorised people;
(b) portable or mobile equipment must be of an explosion-protected design, eg hand torches, vehicles or cleaning machines;
(c) fixed equipment should be of an explosion-protected design. This can also be useful for the purposes of audit or later plant modifications.

388 If the risk assessment indicates that, after appropriate measures have been taken to eliminate or control risks, there is still a significant risk then signs should be used to reduce the risk further. The sign should:

Guidance regulation 7(3) and Schedule 4

(a) be visible to the public and outward-facing on the boundary and fire service entry points;
(b) be positioned at the points of entry to the place where the explosive atmosphere may exist within a clearly defined area;
(c) supplement, if necessary, other means used to identify less clearly defined hazardous zones, eg painted lines on the ground around large open-air plant.

389 Where signage is considered necessary, in addition to other risk reduction measures, it may sometimes be more appropriate just to mark points of entry to the workplace as a whole, rather than the numerous individual locations within the site, if all the special precautions apply throughout the entire site.

390 Signs should be of sufficient size to fulfil their warning function, and they should be maintained so that they are clearly visible. The arrangements made by employers under regulation 9 should ensure that employees receive sufficient information, instruction and training on the meaning of the sign and the measures to be taken in connection with it.

ACOP regulation 7(4)

Verification of places containing explosive atmospheres

391 Verification of explosion safety should be carried out by someone with sufficient training and experience in explosion protection. If sufficient verification expertise exists in-house then it may be performed by that person(s) but the appointed person(s) should be impartial and sufficiently independent to carry out the assessment/verification effectively. Otherwise external competent verification should be obtained. The procedure should confirm that the plant, equipment, protective systems, safety devices, components and their combinations, and the building/structure housing them, are suitable for use with the dangerous substances that are to be used in the workplace and the classification(s) of potentially explosive atmosphere that may exist within this.

Guidance regulation 7(4)

392 The employer must ensure that verification is carried out by someone who is competent to consider the risks at that workplace and decide on the adequacy of the control and other measures to ensure explosion safety. The verification can be in-house as long as the competent person is sufficiently independent of the installation and operation of the work area involved. On larger installations more than one person may need to be involved.

393 The verifier must have obtained sufficient practical and theoretical knowledge from actual experience and/or professional training relevant to the particular workplace and work activity they intend to verify. For example, someone who is competent to verify the explosion safety of a petrol station may not be competent to verify the adequacy of measures to deal with combustible dust in a textile factory.

394 The purpose of verifying overall explosion safety is to confirm the workplace can operate in accordance with these Regulations. The workplace should not be brought into use if verification shows that explosion risks are such that it is not safe to do so. Verification should include consideration of the following:

(a) the dangerous substances that will be present at the workplace, including their hazardous properties and quantity;
(b) the suitability of the plant, equipment and protective systems for work in explosive atmospheres;
(c) the work processes, operating procedures and systems of work;
(d) the effectiveness of measures to:

Guidance regulation 7(4)

- (i) prevent explosive atmospheres forming;
- (ii) control risks from explosive atmospheres;
- (iii) mitigate the effects of an explosion;

(e) the effectiveness of emergency arrangements, where these are required.

395 Verification can be carried out through a variety of means, eg by examination of documents, visual inspection, or physical checks and measurements. Much of the work may be a normal part of the commissioning process. Examples of the work involved include:

(a) checks that mechanical ventilation systems produce the air flows intended;
(b) inspection of records showing that process equipment is leak-tight before dangerous substances are introduced for the first time;
(c) ensuring that a hazardous area classification drawing has been prepared, and a visual inspection that electrical equipment is of the correct type or category for the zone where it has been installed and has been installed correctly;
(d) ensuring that appropriate information is available about the dangerous properties of materials to be handled in the plant.

396 Verification may be used to demonstrate that recommendations of the risk assessment have been put into effect. Where a workplace is subject to legislation on major hazards, such as COMAH, the verification may be used as part of the ALARP argument required by those Regulations.

397 Although there is no requirement to keep a record of the verification, it is recommended that the verifier's name and the date on which verification was completed be recorded. If a record is kept it should be in an easily accessible format. Any documentation produced may be useful for the risk assessment, especially where verification indicates that specific conditions need to be maintained to ensure explosion safety. These conditions should be included in the risk assessment record.

398 Regulation 5(3) requires any risk assessment to be kept under review, for example when new processes are started, or new work equipment is brought into use. Part of that review might involve verification of a significant new plant or process, but it is not intended that verification be applied to every change in a work activity or equipment.

ACOP regulation 7(5)

Provision of antistatic clothing and footwear

399 Where the risk assessment under regulation 5 indicates that an electrostatic discharge could ignite any potential explosive atmosphere that may exist, the employer should:

(a) provide employees working in hazardous places with antistatic ordinary work clothing, including footwear;
(b) where the risk assessment shows this to be necessary, also provide employees with other antistatic work clothing, including any personal protective equipment provided for other purposes.

Guidance regulation 7(5)

400 Some clothing, including footwear, contains materials that can generate electrostatic discharges during use. Such discharges can ignite certain types of explosive atmospheres. Employees should be reminded to also consider the electrostatic properties of their own clothing which may be worn under the protective clothing provided by their employer. Antistatic or ordinary clothing should not be fastened up or removed in places where an explosive atmosphere may occur and a safe area should be established where workers are able to remove or change clothing etc in safety.

Guidance regulation 7(5)

401 The risk from electrostatic discharges from clothing can be reduced if the wearer is earthed by means of suitable footwear and flooring, such as concrete or steel grids. This is likely to be sufficient for places classified into zone 0, 1 or 2 as specified in Schedule 2. In a small number of cases special footwear may also be necessary for zones 20, 21 and 22. Antistatic footwear and flooring should be tested routinely and replaced if it is found that its antistatic properties have deteriorated.

402 Where employee or visitor possessions have the potential to create electrostatic discharges, the employer should have a policy to manage them.

403 General advice on electrostatic hazards is contained in BS 5958-1, parts 1[33] and 2[65] and includes a range of measures applicable to various industrial situations such as petrochemical installations and flammable powder handling. Further information can also be found in *Electrostatics. Code of Practice for the avoidance of hazards due to static electricity*.[34] See also ACOP and guidance on control and avoidance of ignition sources in paragraphs 238–254.

404 This regulation is subject to the transitional provisions in regulation 17(1) to (3). This period has now expired; however, regulation 17(1) still applies to exempt equipment and protective systems that were in operation in workplaces before 30 June 2003 as per regulation 7(3), Schedule 3.

Regulation 8 Arrangements to deal with accidents, incidents and emergencies

Summary of regulation 8

This regulation requires employers to protect the safety of employees by forward planning to have in place arrangements to deal with accidents, incidents and emergencies, including the evacuation, escape or rescue of people. The arrangements include first aid, safety drills and testing, information on hazards, warning and response systems and means of escape. These are required unless such measures would be disproportionate to the risk.

Regulation 8

(1) Subject to paragraph (4), in order to protect the safety of his employees from an accident, incident or emergency related to the presence of a dangerous substance at the workplace, the employer shall ensure that –

(a) procedures, including the provision of appropriate first-aid facilities and relevant safety drills (which shall be tested at regular intervals), have been prepared which can be put into effect when such an event occurs;
(b) information on emergency arrangements, including –
  (i) details of relevant work hazards and hazard identification arrangements; and
  (ii) specific hazards likely to arise at the time of an accident, incident or emergency;

is available;

(c) suitable warning and other communication systems are established to enable an appropriate response, including remedial actions and rescue operations, to be made immediately when such an event occurs;
(d) where necessary, before any explosion conditions are reached, visual, or audible, warnings are given and employees withdrawn; and
(e) where the risk assessment indicates it is necessary, escape facilities are provided and maintained to ensure that, in the event of danger, employees can leave endangered places promptly and safely.

Regulation 8

(2) Subject to paragraph (4), the employer shall ensure that information on the matters referred to in paragraph (1)(a), (c) to (e) and the information required by paragraph 1(b) is –

- *(a) made available to relevant accident and emergency services to enable those services, whether internal or external to the workplace, to prepare their own response procedures and precautionary measures; and*
- *(b) displayed at the workplace, unless the results of the risk assessment make this unnecessary.*

(3) Subject to paragraph (4), in the event of an accident, incident or emergency related to the presence of a dangerous substance at the workplace, the employer shall ensure that –

- *(a) immediate steps are taken to –*
  - *(i) mitigate the effects of the event;*
  - *(ii) restore the situation to normal; and*
  - *(iii) inform those of his employees who may be affected; and*
- *(b) only those persons who are essential for the carrying out of repairs and other necessary work are permitted in the affected area and they are provided with –*
  - *(i) appropriate personal protective equipment and protective clothing; and*
  - *(ii) any necessary specialised safety equipment and plant;*

  which shall be used until the situation is restored to normal.

(4) Paragraphs (1) to (3) shall not apply where –

- *(a) the results of the risk assessment show that, because of the quantity of each dangerous substance at the workplace, there is only a slight risk to employees; and*
- *(b) the measures taken by the employer to comply with his duty under regulation 6(1) are sufficient to control that risk.*

Guidance 8

405 In respect of the dangers arising from an accident, incident or emergency involving the dangerous substance, compliance with regulation 8 of DSEAR fulfils most of the requirements of regulation 8 of the Management Regulations. However, the Management Regulations contain additional requirements to nominate competent persons to initiate procedures, for example.

406 Other health and safety at work legislation also include requirements on planning for and responding to emergencies. These include COMAH, the Offshore Installations (Prevention of Fire and Explosion, and Emergency Response) Regulations 1995 (SI 1995/743), and the Radiation (Emergency Preparedness and Public Information) Regulations 2001 (SI 2001/2975). The emergency measures under DSEAR deal with safety risks. Employers will need to consider whether such arrangements are also required to address health risks under COSHH.

407 The assessment of accidents, incidents and emergencies related to the presence of a dangerous substance, and the determination of emergency arrangements to be taken, should be performed at the same time as the requirements of the other regulations above, plus those of any other relevant health, safety and fire legislation. In practice, meeting the requirements in this other legislation (with regard to flammable properties etc) may mean that an employer is largely meeting the requirements in DSEAR.

ACOP 8

408 The employer should consider the conclusions of their risk assessment about the likelihood and scale or magnitude of the predicted effects on people of any foreseeable unplanned event involving dangerous substances on their premises. The resulting emergency arrangements put in place should aim to minimise the impact.

409 These measures should be taken in addition to the measures already required by regulation 6 to eliminate or reduce risk (by design, and safe operation, instruction and training).

Guidance 8

410 Depending on the findings of the risk assessment and measures already taken, employers will need to consider:

(a) what additional first-aid facilities may need to be provided taking into account the likely effects of any incident;
(b) what additional safety drills may need to be developed (and tested). The frequency of practising any such drills will depend on a number of factors including:
   (i) the quantity of dangerous substances on site and the level of risk they present;
   (ii) the size of the workplace and workforce; and
   (iii) the success or otherwise of previous tests.

411 It can be helpful for employers to involve employees and their representatives during the risk assessment process, as they can be a useful source of information. The emergency arrangements will need to be reviewed and, if necessary, revised if circumstances change at the workplace. For example, if there is a significant increase in the use of a dangerous substance or if new work processes lead to the introduction of new substances into the workplace.

412 Following the risk assessment, there is a legal requirement, under the Safety Representatives and Safety Committees Regulations 1977 and the Health and Safety (Consultation with Employees) Regulations 1996, to consult employees or their representatives on the introduction of any measures which may substantially affect their health and safety.

ACOP 8

413 The warning and other communication systems should be appropriate to the level of risk presented by foreseeable accidents, incidents or emergencies and provide sufficient time and information to allow the necessary emergency actions to be carried out. A warning or communication system should:

(a) not require people to remain in the affected area to continue to sound or give the alarm during an emergency;
(b) be rapid, easy and reliable to initiate;
(c) be appropriate to the level of foreseeable risk from an event;
(d) allow enough time and information to take necessary actions; and
(e) be seen or heard in all areas of the workplace likely to be affected by the incident.

Guidance 8

414 Employers should also have procedures in place to keep employees informed of situations as they develop and any actions that may be needed as a consequence.

415 Examples of warning systems include:

(a) a continuous or intermittent ringing bell;
(b) a klaxon or hooter;

Guidance 8

(c) warning lights;
(d) an intercom or tannoy system.

416 When considering what warning and communication systems will be appropriate, employers should take into account:

(a) the size of their workplace and workforce;
(b) who needs to be alerted and why;
(c) quantities of substances involved and the level and type of risk those substances present; and
(d) the emergency actions to be taken in the event of an incident and the required response times for these.

417 Warning systems are not necessarily restricted to signalling the need for withdrawal or evacuation of people who might be affected by the incident. They can also be to alert employees of an incident or emergency, so that they can take appropriate emergency action to contain or mitigate the incident. Where there are multiple alarm systems in a workplace, employers should ensure that these are clearly discernible and that their employees have the necessary training and equipment to be able to safely carry out the correct actions required.

ACOP 8

418 Early detection and warning measures should be in place if there is the potential for a release of dangerous substance of sufficient quantity to create a significant explosion hazard. Where appropriate, immediate measures should be in place to detect such a release before it reaches its LEL.

419 The employer is responsible for selecting a suitable place or places of safety where people can gather and be accounted for after vacating their workplaces.

Guidance 8

420 Employers should already have considered mitigation measures against fire and explosion by the provision of adequate escape facilities in parallel with the requirement contained in general fire safety legislation (see paragraphs 20–22). The installation and type of warning system is dependent on the nature of the operation, eg an early detection system would be required in an aerosol filling shed but not in a small storage site.

421 However, the presence of a dangerous substance can accelerate the spread of fire and production of smoke and other toxic fumes. Particular attention should be given in choosing the escape route and place of safety to ensure that it will not be affected by the event should it escalate. Employers will also need to take account of situations where toxic substances may be released as a result of an incident. When considering escape facilities employers should assess the potential for explosions, rapid fire development and ingress of dangerous substances into escape routes which may compromise escape.

ACOP 8

422 The employer should ensure sufficient information on the nature of any foreseeable emergencies involving dangerous substances is made available to the relevant accident and emergency services who are likely to be asked to deal with such incidents. As a minimum, employers should contact the external emergency services, inform them that this information is available and offer to send it to them. Employers should also make this information available to any on-site emergency services.

Guidance 8

423 Full mitigation of foreseeable accidents, incidents and emergencies will typically be achieved by a combination of workplace emergency arrangements and those provided by the emergency services overall to ensure the safety of employees, emergency service personnel and other people. Employers will need to

Guidance 8

consider if some or all of the external emergency services need to be aware of their emergency arrangements. In the case of offshore installations, the coastguard may also need to be included. The fire service will in any case assume responsibility for tackling any fire upon their arrival, but they may also be able to assist in dealing with other non-fire emergencies such as released or spilled dangerous substances.

424 If requested by the emergency services, employers should be prepared to send the information to them and/or meet with their representatives to discuss the emergency procedures. The information made available will help the emergency services to prepare their own response procedures and should include:

(a) the identity, location and approximate quantities of dangerous substances;
(b) the foreseeable types of accident, incident or emergency that could occur and the hazards that may result;
(c) where on site such events could occur, what effects they could have, other areas that may be affected should the event escalate and the possible repercussion that may cause;
(d) the emergency arrangements drawn up by the employer to deal with accidents, incidents and emergencies, the procedures prepared by the employer to deal with any such event, the warnings and other communication systems, and escape facilities; and
(e) details of the employer's designated contact who will advise the emergency services of the situation on their arrival at an emergency.

425 Information, training and instruction on emergency arrangements should be made available to employees and their representatives. For example, this may be by the periodic circulation of copies of the arrangements, or providing individual copies. The adequacy of emergency procedures should be regularly tested and any deficiencies found should be addressed. Unless the risk assessment indicates otherwise, employers should display the emergency procedures in a prominent position at key locations in the workplace. Appropriate information on emergency arrangements should also be communicated to non-employees who may be affected.

426 Employers should keep the emergency services updated on significant changes, such as the quantities and nature of dangerous substances present.

ACOP 8

427 Employers should implement those measures necessary to achieve control or containment of an accident, incident or emergency to allow sufficient time for people to escape or be evacuated to a place of safety. Following an event, employers must assess whether any danger remains and carry out the necessary measures to make the situation safe. If there are any doubts about safety, expert assistance should be sought, eg from accident and emergency services. Implementation of the necessary measures should be achieved without exposing employees or others to unnecessary risk, giving precedence to remote control measures over those requiring entry into the affected area.

Guidance 8

428 Employers should adopt a systematic approach for identifying potential accidents, incidents, emergencies or other events and consider how they can be detected when they have occurred or are occurring. They also need to determine, for the various stages of the accident, incident or emergency, the appropriate type and degree of intervention to both mitigate the consequences of the event and prevent its further escalation. The impact of an accident, incident or emergency can often be greatly reduced if prompt and correct action is taken as soon as the event occurs.

Guidance 8

429 Factors to be taken into account by employers when assessing the requirements for emergency arrangements include:

(a) the properties of the dangerous substances present, their quantities and the way they are used or stored;
(b) the foreseeable types of accidents, incidents, emergencies or other events that may occur, and the level of risk that may be presented (eg the response required to deal with a major fire in a bulk storage facility will be different from that required for a small spill of a few litres of flammable liquid);
(c) precursors to the end emergency (fire or explosion) involving the dangerous substance (eg unignited leaks, spills and releases of the dangerous substance, or the potential for these, arising for example, from mechanical damage to plant containing dangerous substances);
(d) means of detecting events, eg the selection and effectiveness of the means of detecting a leak, spill or release of a dangerous substance and hence the speed and nature of the emergency response will depend on such matters as location, size of release and potential escalating events;
(e) the trigger events for alarms and warnings, eg where mechanical exhaust ventilation is provided to ensure a safe atmosphere, it might well be considered reasonable that critical reduction in its flow should cause an alarm or suitable warning to be given, so that emergency actions can be taken to isolate the release of the dangerous substance and take other remedial action as necessary;
(f) the role of non-employees in emergency arrangements, eg a spill outdoors during a driver-controlled tanker unloading operation. In this case it would be reasonable for the driver (who may not be an employee) to detect the spill and initiate appropriate emergency action which might include warning other people in the locality, using spill control equipment and calling the emergency services;
(g) specific procedures that employees and others should follow if an accident, incident or emergency occurs (eg clearing up spills of flammable liquids or, for more serious incidents, moving to a safe area or complete evacuation of the workplace);
(h) the role, responsibilities and authority of employees who may be allocated specific duties (eg people responsible for shutting down equipment, checking that specific areas have been successfully evacuated, contacting the emergency services etc);
(i) the provision, where necessary, of suitable safety equipment or personal protective equipment; and
(j) procedures for assisting particular groups of people, such as members of the public or other visitors on site (who may be unfamiliar with the workplace and the risks from dangerous substances that are present) or disabled employees.

430 Steps to mitigate the effects of an incident may, where it is safe to do so, include:

(a) evacuating people who may be affected, taking into account possible escalation of the incident, to a place of safety;
(b) isolating plant or equipment from where uncontrolled releases of a dangerous substance are occurring;
(c) removing the dangerous substance under threat to a safe place;
(d) preventing the further spread of a spilt or leaking dangerous substance by the use of barriers, booms or absorbent materials;
(e) limiting the extent of any flammable vapour cloud arising from a release of the dangerous substance by, for example, the use of water sprays and curtains, or applying firefighting foam over the surface of the spilt or leaking liquid materials;

Guidance 8

(f) increasing natural or mechanical ventilation to dilute hazardous concentrations of dangerous substances arising from an incident;
(g) controlling potential ignition sources in non-hazardous areas that are now affected by an uncontrolled release of dangerous substance;
(h) protecting the vessels or plant containing the dangerous substance against the effects of fire by such means as water deluge systems, water monitors and passive fire protection coatings; and
(i) applying appropriate firefighting materials to a fire involving a dangerous substance.

431 Under regulation 8, measures to restore the situation to normal following an event are limited to those measures needed to achieve the normal level of safety for the premises. They do not include measures to rebuild a plant or restore it to normal production or operation but could include:

(a) repairing or decommissioning leaking or unsafe plant;
(b) safe recovery and clean-up of spilt or leaked dangerous substances;
(c) making safe damaged or unstable buildings;
(d) repair or replacement of any equipment, monitoring devices or alarms necessary for the safety of employers or others present on the premises;
(e) neutralising or disposing of any unstable or dangerous substances resulting from an incident.

432 Measures taken to deal with accidents, incidents and emergencies will need to be adaptable to deal with the specific situation. This is also the case with the remedial action following the accident, incident or emergency. Employers must therefore ensure that those of their employees expected to respond in the event of an accident, incident or emergency involving a dangerous substance have the necessary skills, expertise and training to carry out the functions expected of them; and that suitable plant and equipment necessary to carry out these functions is available and properly maintained for immediate use.

433 Equipment to be provided should include any PPE (including appropriate protective clothing and footwear) that is necessary for employees to safely carry out the emergency actions required. However, employees should not be exposed to unnecessary risk in carrying these out. When selecting equipment to be used in emergencies, employers should ensure it is appropriate for the circumstances in which it may be used. For example, the need to avoid ignition sources when dealing with the releases of flammable vapours, gases etc.

434 Before carrying out any remedial work the employer should carry out a risk assessment to determine the control measures that need to be put in place to ensure the health and safety of employees. As part of this assessment the employer should determine the appropriate systems of work, including possible permit-to-work systems that need to be implemented before employees or contractors enter any areas affected by the incident or emergency.

ACOP 8

435 Employers are not required to implement additional emergency arrangements if they have determined through the risk assessment carried out under regulation 5(1) that the risk is slight because of:

(a) the quantity of each substance present;
(b) the effectiveness of existing emergency arrangements;
(c) control measures in place that fulfil the requirements of regulation 6 and other fire safety legislation.

Guidance 8

436 In respect of fire, the normal emergency procedures (including escape routes and means of giving warning) already provided in the workplace may be assessed to be sufficient. General guidance on fire precautions is available on HSE's website at www.hse.gov.uk/toolbox/fire.htm (also see paragraph 90). Employers must be satisfied, however, that those control measures will continue to be effective if an accident, incident or emergency occurs. For example, where an accompanying event such as an explosion is possible that might compromise an escape route, alternative measures may need to be considered to ensure safe evacuation. For example, this might include a specially strengthened escape route and/or refuge. The escape route from a non-hazardous area must not be through a hazardous area.

Regulation 9 Information, instruction and training

Summary of regulation 9

Appropriate information, training and instruction should be given to contractors and employees on the dangerous substances present together with information on the hazards, risks, precautions and actions necessary for them to remain safe.

Regulation 9

(1) Where a dangerous substance is present at the workplace, the employer shall provide his employees with –

(a) suitable and sufficient information, instruction and training on the appropriate precautions and actions to be taken by the employee in order to safeguard himself and other employees at the workplace;
(b) the details of any such substance including –
(i) the name of the substance and the risk which it presents;
(ii) access to any relevant safety data sheet; and
(iii) legislative provisions which concern the hazardous properties of the substance;
and
(c) the significant findings of the risk assessment.

(2) The information, instruction and training required by paragraph (1) shall be –

(a) adapted to take account of significant changes in the type of work carried out or methods of work used by the employer; and
(b) provided in a manner appropriate to the risk assessment.

ACOP 9

437 The information provided to employees as required under regulation 4(1)(b) and, to the extent that it is required by the nature and degree of the risk, to other people who may be present at a workplace, should include the following:

(a) how and where the dangerous substance is used in the specific site activities in addition to the general information in the SDS;
(b) the precautions and actions mentioned in regulation 9(1) as part of the information for employees include the control and mitigation measures adopted, including methods of work, the reasons behind them, and how to use them properly;
(c) training and instruction, which should include the reasoning (theory) behind the practice. Training in the use and application of control measures and equipment should be carried out taking into account recommendations and instructions supplied by the manufacturer;

ACOP 9

(d) any procedures for dealing with accidents, emergencies and incidents prepared in accordance with regulation 8. This ranges from smaller unplanned incidents (including dealing with faults, clearing blockages) to larger emergencies and should prepare staff for how to react if and when foreseeable events happen;
(e) any further relevant information resulting from a review of the risk assessment, why it has been done and how any changes will affect the way employees do the work in the future.

438 The information, instruction and training provided should be appropriate to the level of understanding and experience of employees. It should be provided in a form which takes account of any language difficulties or disabilities. Information can be provided in whatever form is most suitable in the circumstances, as long as it can be understood by everyone.

439 Should change to any work processes occur, then the employer has a duty to ensure extra information, instruction and training is given if necessary.

Guidance 9

440 The control measures necessary for the safe handling and use of dangerous substances often require or are dependent on employees carrying out the appropriate operating procedures correctly and complying with written or verbal instructions. Employers, therefore, should provide employees with sufficient supervision and training to ensure that the systems of work required by regulation 6 and Schedule 1 are fully implemented and operating procedures are correctly followed.

441 The objective of providing information, instruction and training is to ensure that employees can work with dangerous substances without putting themselves or others at risk. The extent of the information, instruction and training required will be proportionate to the degree of complexity of the hazards, risks, processes and controls.

442 The significant findings of the risk assessment will help to explain to employees what the risks are and how the control/mitigation measures are designed to protect their safety. It will also help employees to understand and use the safeguards that employers introduce.

443 The employer should consider all the various ways of providing information, instruction and training and select those most appropriate to their own circumstances. Options include:

(a) class or group tuition;
(b) individual tuition;
(c) written instructions including leaflets, courses etc;
(d) refresher training, toolbox talks etc.

444 Employers also need to take account of the needs of people other than employees who may be present on site, such as contractors, members of the public etc. While it may not always be practical to provide formal training in these circumstances, employers should consider what other information or instruction may be needed to reduce risks. For example, pictorial signs for infrequent visitors to the site or those for whom English is not their first language (which might be the case for delivery drivers and other staff/visitors), notices explaining hazards (eg warning notices, no smoking signs etc), and copies of emergency and evacuation procedures.

Guidance 9

445 For employees with little or no understanding of English, or those who cannot read English, employers may need to make special arrangements. These could include providing translation, using interpreters, or replacing written notes with clearly understood symbols or diagrams.

446 Where employees from one employer work on the premises of another, the employer occupying the premises must provide the other employer with sufficient information about any dangerous substances that may be present at the premises as part of the day-to-day activity. This information should be sufficiently detailed to allow the other employer to provide his own employees with information and any appropriate instruction on complying with the occupying employer's measures.

447 The employer occupying the premises will also need to know about any dangerous substances that are likely to be used or produced by the work the other employer will be doing. This information is important as it allows the occupying employer to:

(a) be satisfied that the measures put in place by the employer doing the work will not only protect their own employees from risks presented by the substances concerned, but also the occupier's own employees;
(b) provide their own employees with information and instruction about any dangerous substances that the other employer will be using or the work will produce;
(c) reassure their employees that any risks to their safety are being properly controlled; and
(d) take steps to ensure that emergency services personnel attending in the event of an incident are made aware of dangerous substances on the premises posing significant risk to their safety.

448 If changes to workplace activities necessitate a revised risk assessment, then it may be necessary for employers to provide updated information, instruction or training to employees (see also paragraph 147).

449 Information, instruction and training in relation to risks from dangerous substances need only be provided to non-employees where it is required to ensure their safety. For example, it would not be required in situations where the number and type of visitors to a workplace, or the short duration of the visit, combined with a negligible risk, make the provision of such information inappropriate (such as customers in a shop). However, where it is provided, it should be in proportion to the level and type of risk.

Regulation 10 Identification of hazardous contents of containers and pipes

Summary of regulation 10 and Schedule 5

This regulation applies to containers and pipes that contain dangerous substances that are not subject to or are exempt from any marking requirements in the legislation listed in Schedule 5.

Regulation 10

Where containers and pipes used at work for dangerous substances are not marked in accordance with relevant requirements of the legislation listed in Schedule 5, the employer shall, subject to any derogations provided for in that legislation, ensure that the contents of those containers and pipes, together with the nature of those contents and any associated hazards, are clearly identifiable.

Schedule regulation 5 to 10

Schedule 5 Legislation concerned with the marking of containers and pipes

The Health and Safety (Safety Signs and Signals) Regulations 1996 (S.I. 1996/341);

The Good Laboratory Practice Regulations 1999 (S.I. 1999/3106);

The Chemicals (Hazard Information and Packaging for Supply) Regulations 2009 (S.I. 2009/716);

The Carriage of Dangerous Goods and Use of Transportable Pressure Equipment Regulations 2009 (S.I. 2009/1348).

ACOP 10

450 In many cases the marking of the majority of plant and equipment in which dangerous substances are conveyed and stored, including their location, is already covered by other legislation. Regulation 10 does not require everything to be marked or labelled, but the employer should decide, through their risk assessment, if and how contents of containers and pipes containing dangerous substances should be identified, whether appropriate identification is required and if so, the form it should take.

Guidance 10

451 Identification is not necessary where the substance is a bulk solid product, such as flour, which is not itself a dangerous substance and is only hazardous if released from containment and dispersed in the air.

452 Identification of pipes and containers:

(a) alerts people to the presence of a dangerous substance so that they can take the necessary precautions;
(b) can also help to avoid confusion over contents and thereby avoid incorrect mixing of contents.

453 A container includes any fixed or portable, open or enclosed means to contain dangerous substances such as tanks, silos, reaction vessels and waste receptacles together with any associated pipe runs or piping system.

454 The regulation allows a common-sense approach to selecting means of identification, which will depend on the work activity and take into account security implications. Suitable means could include labelling, the use of appropriate colour coding, or instructions and training.

455 In situations where the contents may change regularly, eg chemical process vessels and pipes which are not dedicated to one substance, test tubes in laboratories etc, labelling may not be practicable. In these cases employers will need other arrangements to highlight to employees the hazards associated with the substances involved. Employers could provide process instruction sheets, record sheets or training for employees to alert them to the risk and the need for caution.

Regulation 11 Duty of co-ordination

Summary of regulation 11

Regulation 11 addresses the need to co-ordinate explosion protection measures where employers share the same workplace.

Regulation 11

Where two or more employers share the same workplace (whether on a temporary or a permanent basis) where an explosive atmosphere may occur, the employer responsible for the workplace shall co-ordinate the implementation of all the measures required by these Regulations to be taken to protect employees from any risk from the explosive atmosphere.

ACOP 11

456 The employer responsible for the workplace should record as required by regulation 5(4) the 'aim of co-ordination' in the risk assessment as well as the measures and procedures for implementing it.

Guidance 11

457 The aim of co-ordination is to safeguard people by:

(a) alerting other employers, employees and others at the workplace to the potential presence of hazardous places;
(b) ensuring that suitable control and mitigation measures are in place;
(c) ensuring employees and others have sufficient training etc (see regulation 9); and
(d) facilitating emergency arrangements in the event of an incident.

458 In shared premises, it may be the employer responsible for the workplace or one of the other employers who shares the premises whose work activity creates the explosive atmosphere. Effective co-ordination will require the responsible employer for the workplace to obtain sufficient information from all employers at the workplace about the nature of their work activity, whether any dangerous substances are likely to be present, and the likelihood of an explosive atmosphere occurring. All employers at the workplace should provide the responsible employer with the information required and assist in assessing the shared risks and implementing any necessary safety measures.

459 In multi-occupancy premises, it is not expected that an explosive atmosphere would extend into shared common areas. However, the occupier/employer should still carefully consider the potential for and effect of an incident involving dangerous substances on the rest of the premises including the common parts and liaise with the other occupants to ensure adequate emergency procedures are in place (see Section 3 of the HSW Act).

Regulations 12–16 deal with the application of DSEAR outside Great Britain, exemptions from the Regulations, and amendments to and removal of other legislation.

Regulation 12 Extension outside Great Britain

Regulation 12

These Regulations shall apply outside Great Britain as sections 1 to 59 and 80 to 82 of the 1974 Act apply by virtue of the Health and Safety at Work etc. Act 1974 (Application outside Great Britain) Order 2013.

Regulation 13 Exemption certificates

Regulation 13

(1) Subject to paragraph (2), the Health and Safety Executive may, by a certificate in writing, exempt any person or class of persons or any dangerous substance or class of dangerous substances from all or any of the requirements or prohibitions imposed by or under these Regulations and any such exemption may be granted subject to conditions and to a limit of time and may be revoked at any time by a certificate in writing.

(2) The Health and Safety Executive shall not grant any such exemption unless, having regard to the circumstances of the case, and in particular to –

(a) the conditions, if any, which it proposes to attach to the exemption; and
(b) any requirements imposed by or under any enactments which apply to the case,

it is satisfied that the health and safety of persons who are likely to be affected by the exemption will not be prejudiced in consequence of it and that the exemption will be compatible with the requirements of the Directives.

(3) For the purposes of paragraph (2), "the Directives" means Council Directive 98/24/EC on the protection of the health and safety of workers from the risks related to chemical agents at work and Council Directive 99/92/EC on minimum requirements for improving the safety and health protection of workers potentially at risk from explosive atmospheres.

Regulation 14 Exemptions for Ministry of Defence etc

Regulation 14

(1) In this regulation –

(a) "Her Majesty's Forces" means any of the naval, military or air forces of the Crown, whether raised inside or outside the United Kingdom and whether any such force is a regular, auxiliary or reserve force, and includes any civilian employed by those forces;
(b) "visiting force" has the same meaning as it does for the purposes of any provision of Part 1 of the Visiting Forces Act 1952; and
(c) "headquarters" means a headquarters for the time being specified in Schedule 2 to the Visiting Forces and International Headquarters (Application of Law) Order 1999.

(2) The Secretary of State for Defence may, in the interests of national security, by a certificate in writing, exempt –

(a) any of Her Majesty's Forces;
(b) any visiting force;
(c) any member of a visiting force working in or attached to a headquarters; or
(d) any person engaged in work involving dangerous substances, if that person is under the direct supervision of a representative of the Secretary of State for Defence;

from all or any of the requirements or prohibitions imposed by these Regulations and any such exemption may be granted subject to conditions and to a limit of time and may be revoked at any time by a certificate in writing, except that, where any such exemption is granted, suitable arrangements shall be made for the assessment of the risk to safety created by the work involving dangerous substances and for adequately controlling the risk to persons to whom the exemption relates.

Regulation 15 Amendments

Regulation 15

(1) The Acts and instruments referred to in Part 1 of Schedule 6 shall be amended in accordance with that Part.

(2) The instruments referred to in Part 2 of Schedule 6 shall be amended in accordance with that Part.

Guidance 15

460 This regulation brings Schedule 6 into force. It contains amendments to legislation (mainly concerning petrol).

461 Schedule 6 is not reproduced here because much of it is out of date or is likely to be to be changed by repeals or modifications linked to a 2013 review of petroleum legislation. The original text is readily accessible by searching for the DSEAR statutory instrument SI 2002/2776 on www.legislation.gov.uk.

Regulation 16 Repeals and revocations

Regulation 16

(1) The Act and instruments referred to in column 1 of Part 1 of Schedule 7 shall be repealed or revoked to the extent specified in the corresponding entry in column 3 of that Part.

(2) The Act and instruments referred to in column 1 of Part 2 of Schedule 7 shall be repealed or revoked to the extent specified in the corresponding entry in column 3 of that Part.

Schedule 7

Schedule 7 Repeals and revocations

Part 1 Repeal and revocation

Column 1 ***Title***	**Column 2** ***Reference***	**Column 3** ***Extent of repeal or revocation***
The Petroleum (Consolidation) Act 1928	*c 32*	*The proviso to section 2(1).* *Section 9.* *Section 17.* *In section 25A, the word "and" at the end of paragraph (a).*
The Celluloid, etc Factories, and Workshops Regulations 1921	*SR & O 1921/1825*	*The whole Regulations.*
The Manufacture of Cinematograph Film Regulations 1928	*SR & O 1928/82*	*The whole Regulations.*

Schedule 7

Part 1 Repeal and revocation

Column 1 *Title*	Column 2 *Reference*	Column 3 *Extent of repeal or revocation*
The Petroleum (Carbide of Calcium) Order 1929	*SR & O 1929/992*	*The whole Order.*
The Petroleum (Compressed Gases) Order 1930	*SR & O 1930/34*	*The whole Order.*
The Cinematograph Film Stripping Regulations 1939	*SR & O 1939/571*	*The whole Regulations.*
The Petroleum (Carbide of Calcium) Order 1947	*SR & O 1947/1442*	*The whole Order.*

Part 2 Repeal and revocation

Column 1 *Title*	Column 2 *Reference*	Column 3 *Extent of repeal or revocation*
The Factories Act 1961	*c 34*	*Section 31.*
The Magnesium (Grinding of Castings and other Articles) Special Regulations 1946	*SR & O 1946/2197*	*The whole Regulations.*
The Dry Cleaning Special Regulations 1949	*SI 1949/2224*	*The whole Regulations.*
The Factories (Testing of Aircraft Engines and Accessories) Special Regulations 1952	*SI 1952/1689*	*The whole Regulations.*
The Shipbuilding and Ship-repairing Regulations 1960	*SI 1960/1932*	*Regulations 48 to 52, 54 and 55 to 66.*
The Highly Flammable Liquids and Liquefied Petroleum Gases Regulations 1972	*SI 1972/917*	*The whole Regulations.*
The Abstract of Special Regulations (Highly Flammable Liquids and Liquefied Petroleum Gases) Order 1974	*SI 1974/1587*	*The whole Order.*
The Dry Cleaning (Metrication) Regulations 1983	*SI 1983/977*	*The whole Regulations.*

Schedule 7

Part 2 Repeal and revocation

Column 1 *Title*	**Column 2** *Reference*	**Column 3** *Extent of repeal or revocation*
The Factories (Testing of Aircraft Engines and Accessories) (Metrication) Regulations 1983	*SI 1983/979*	*The whole Regulations.*
The Dangerous Substances in Harbour Area Regulations 1987	*SI 1987/37*	*Regulation 29(a).*
The Workplace (Health, Safety and Welfare) Regulations 1992	*SI 1992/3004*	*Regulation 6(3)(b).*
The Carriage of Dangerous Goods (Classification, Packaging and Labelling) and Use of Transportable Pressure Receptacles Regulations 1996	*SI 1996/2092*	*Regulation 22(b).*
The Carriage of Dangerous Goods by Road Regulations 1996	*SI 1996/2095*	*Schedule 12.*

Regulation 17 Transitional provisions

Summary of regulation 17

Regulation 17 sets out the transitional arrangements for workplaces and work equipment where explosive atmospheres may occur. The period of transition has now elapsed, the text of this regulation has therefore not been reproduced in this publication. Historical information on ensuring compliance in the workplace may be found in Appendix 6.

Appendix 1 Notice of Approval

By virtue of section 16(4) of the Health and Safety at Work etc Act, and with the consent of the Secretary of State for Work and Pensions, the Health and Safety Executive has on 30 October 2013 approved the revised Code of Practice *Dangerous Substances and Explosive Atmospheres: Dangerous Substances and Explosive Atmospheres Regulations 2002. Approved Code of Practice and guidance* (Second edition, 2013, L138).

The revised Code of Practice gives practical guidance on the Dangerous Substances and Explosive Atmospheres Regulations 2002.

By virtue of section 16(5) and with the consent of the Secretary of State for Work and Pensions under that paragraph, the Health and Safety Executive has withdrawn its approval of the following Codes of Practice which shall cease to have effect on 9 December 2013:

- L134 *Design of plant, equipment and workplaces. Dangerous Substances and Explosive Atmospheres Regulations 2002. Approved Code of Practice and guidance*
- L135 *Storage of dangerous substances. Dangerous Substances and Explosive Atmospheres Regulations 2002. Approved Code of Practice and guidance*
- L136 *Control and mitigation measures. Dangerous Substances and Explosive Atmospheres Regulations 2002. Approved Code of Practice and guidance*
- L137 *Safe maintenance, repair and cleaning procedures. Dangerous Substances and Explosive Atmospheres Regulations 2002. Approved Code of Practice and guidance*
- L138 *Dangerous Substances and Explosive Atmospheres. Dangerous Substances and Explosive Atmospheres Regulations 2002. Approved Code of Practice and guidance* (2003 edition)

This Code of Practice comes into effect on 9 December 2013.

Signed

TERESA QUINN
Secretary to the Board of the Health and Safety Executive

3 December 2013

Appendix 2 Other relevant legislation

Relationship with other health and safety legislation

1 The duties in DSEAR apply alongside the HSW Act, other regulations made under the Act, and legislation on fire precautions and within a wider legislative context. Because of their close relationship with DSEAR, an introduction to general fire safety legislation and the EPS regulations is included earlier in this document (see paragraphs 20–26).

The Management of Health and Safety at Work Regulations 1999

2 The Management of Health and Safety at Work Regulations (SI 1999/3242) (the Management Regulations) also require employers and the self-employed to assess the general risks to health and safety arising from their work activity and identify the preventive and protective measures that need to be taken to control the identified risks.

3 The more specific provisions of DSEAR will only apply where dangerous substances are present or used. For example, an assessment of the risks from dangerous substances and arrangements for emergencies carried out under DSEAR will not need to be repeated for the Management Regulations, and in many cases may be incorporated into the more general 'management' assessment as a subset. There are legal requirements, in other regulations, to consult employees. It can be helpful to involve employees or their representatives when carrying out risk assessments.

The Control of Substances Hazardous to Health Regulations 2002

4 Most substances that may be dangerous (to safety) in terms of DSEAR will also present a health risk for which the Control of Substances Hazardous to Health Regulations 2002 (SI 2002/2677) (COSHH) will apply. Employers will have duties to control the risks from those substances under both sets of regulations but the solutions for both are likely to be common.

Classification, labelling and packaging of dangerous substances

5 In 2013/14 the area of classification packaging and labelling is in transition. These developments were driven by the European Regulation (EC) No 1272/2008 on Classification, Labelling and Packaging of Substances and Mixtures (known as the CLP Regulation or CLP). CLP is a direct-acting regulation.

6 By 2015 CLP will replace the existing European system on the classification, labelling and packaging of chemicals – the Dangerous Substances Directive (67/548/EEC) and the Dangerous Preparations Directive (99/45/EC). Both these earlier Directives are implemented in Great Britain by the Chemicals (Hazard Information and Packaging for Supply) Regulations 2009 (SI 2009/716) (CHIP). For these purposes CHIP will lapse in 2015. There are transitional periods in CLP but by 1 June 2015, chemical suppliers must comply only with the CLP Regulation.

7 There are implications for DSEAR because regulation 2 refers to the Approved Classification and Labelling Guide and CHIP which will both lapse. The CLP Regulation now contains the list of harmonised classifications in Table 3.2 in part 3 of Annex VI of CLP. These harmonised classifications and accompanying labelling requirements are legally binding on all chemical suppliers. As a consequence, the Approved Supply List is discontinued and should no longer be used. Appendix 3 of this document contains more information.

Registration, Evaluation, Authorisation and Restriction of Chemicals EC no 1907/2006 (REACH)

8 REACH is a direct-acting European regulation. It regulates the placing of substances and articles into the marketplace and affects chemicals supply, packaging and labelling. The legal provisions relating to safety data sheets now appear in Article 31 and Annex II of REACH.

The Provision and Use of Work Equipment Regulations 1998

9 The Provision and Use of Work Equipment Regulations 1998 (SI 1998/2306) (PUWER) require employers, and people in control to provide safe work equipment, to use it safely and to maintain it.

The Personal Protective Equipment Regulations 1992

10 Under the Personal Protective Equipment Regulations 1992 (SI 1992/2966) employers must select, provide and maintain (at no charge) appropriate protective equipment for employees whenever risks are not adequately protected against by other means. Employers should also provide information, training and instruction for employees on its use and maintenance.

The Petroleum (Consolidation) Act 1928 and associated regulations

11 DSEAR amended the workplace application of petrol safety legislation. Site operators that store petrol for dispensing into vehicles, ships, boats and planes must be licensed under petrol safety law but DSEAR also applies to activities at these sites. Where petrol is stored for different (non-dispensing) purposes, such as factories etc, a petrol licence is not required, but compliance with DSEAR is necessary.

12 DSEAR does not apply to activities on sites that are not workplaces; therefore the current legislation concerning the storage of petrol in domestic premises remains in place.

Appendix 3 Revocation of CHIP and the application of the European CLP Regulation

1 From 1 June 2015, European Regulation (EC) No 1272/2008 on Classification, Labelling and Packaging of Substances and Mixtures (known as the CLP Regulation or CLP) will repeal both the Dangerous Substances Directive (67/548/EEC) and the Dangerous Preparations Directive (99/45/EC) and will become the primary legislation in this area. As a consequence the Chemicals (Hazard Information and Packaging for Supply) Regulations 2009 (CHIP), which implement both these directives, will be revoked. One provision of CHIP relating to data retention and dangerous preparations will remain until June 2018, after which CHIP is fully revoked.

2 Across the EU, the CLP Regulation adopts the United Nations' Globally Harmonised System on chemical classification and labelling (GHS). The CLP Regulation is directly acting on all EU member states. This means that in contrast to directives, no national transposition is necessary and the need for the CHIP Regulations falls away.

3 The CLP Regulation applies to substances placed on the market from 1 December 2010. It will apply to mixtures (formerly called preparations) from 1 June 2015. Chemical suppliers can apply CLP to mixtures placed on the market ahead of the mandatory compliance date if they choose to do so (see below).

4 After 1 June 2015, chemical suppliers must comply only with the CLP Regulation.

Implications for DSEAR

5 The migration to the CLP Regulation means that DSEAR employers will need to become familiar with the new classification hazard terminology and accompanying hazard pictograms relevant to explosive atmospheres. The new CLP pictograms and wording will increasingly be seen on SDSs and product labels and this will need to be taken into account for employee information, training and instruction.

What does this mean for chemical users?

6 Chemical users will see new designs, wording and symbols on product hazard labels. The CLP hazard pictograms are very similar to the CHIP hazard symbols but they have a new shape, design and colour.

7 New hazard statements will replace the CHIP risk phrases (R phrases).

8 New precautionary statements will replace the CHIP safety phrases (S phrases).

9 Guidance on complying with CLP is developed and agreed at EU rather than national level and a detailed suite of guidance is available on the web pages of the European Chemicals Agency (ECHA) http://echa.europa.eu.

10 The legal provisions relating to SDSs now appear in Article 31 and Annex II of REACH.

11 Chemical suppliers must classify, label and package their preparations according to CHIP until 1 June 2015 (regulations 4 and 6–11 refer). As an alternative, chemical suppliers can classify, label and package mixtures (preparations) according to the CLP criteria ahead of this date if they choose to do so. In this case, suppliers must, in addition, continue to classify the mixture under regulation 4 of CHIP, although this information should only be included in an SDS (the requirements on labelling and packaging in regulations 6 to 11 of CHIP will no longer apply).

12 During this transitional period, product hazard labels must only show the details of one regime or the other – 'mixed' labels are not permitted. If a mixture is classified according to CHIP, it must only be labelled and packaged according to CHIP. Conversely, if a mixture is classified according to CLP, it must only be labelled and packaged according to CLP.

Withdrawal of the HSE Approved Supply List

13 The Approved Supply List was published in Great Britain and reproduced the list of dangerous substances that had been assigned harmonised classifications and accompanying labelling requirements agreed by all EU Member States and which appeared in Annex I to the Dangerous Substances Directive (67/548/EEC).

14 However, the CLP Regulation repealed Annex I in full. As a consequence, the Approved Supply List was discontinued and should no longer be used. It is out of date and has no legal effect.

15 It was important, however, that the 40 years of scientific development and technical understanding of the most dangerous chemicals should not be lost. The CLP Regulation immediately re-enacted the list of harmonised classifications which now appears in Part 3 of Annex VI of CLP. The list is represented in two Tables:

(a) Table 3.1 provides the classification and labelling information using CLP criteria and terminology.
(b) Table 3.2 provides the same information but uses the Dangerous Substances Directive (CHIP) criteria and terminology. (Table 3.2 will be repealed from 1 June 2015 when CLP applies in full and the 'old' list is no longer needed).

16 Harmonised classifications and the accompanying labelling requirements are legally binding where they are used.

17 Annex VI of CLP is the official list of harmonised classifications and should be the only source for those seeking this information. The Annex is easily searchable by a number of different fields at http://esis.jrc.ec.europa.eu/index.php?PGM=cla.

18 Regulation 14 of CHIP currently sets out the enforcement arrangements for both CHIP and the CLP Regulation. This will continue to be the case until 1 June 2015. The enforcement arrangements for the CLP Regulation will then appear in a new set of regulations. In addition to the enforcement of CLP, these regulations will also include the enforcement arrangements for the biocides regime, 'prior informed consent' (PIC – export of dangerous chemicals) and the appointment of the UK Competent Authorities for CLP, biocides and PIC.

Guidance on the CLP Regulation

19 Employers subject to DSEAR who need advice on the CLP Regulation should look to the suite of detailed guidance supporting the Regulation provided by the European Chemicals Agency (ECHA) http://echa.europa.eu/support. Alternatively, employers can contact the UK CLP Helpdesk: ukreachca@hse.gsi.gov.uk.

20 Another source of helpful information is ECHA's Classification and Labelling Inventory which provides information on all substances placed on the EU market and notified to ECHA: http://echa.europa.eu/en/regulations/clp/cl-inventory.

Appendix 4 Fire resistance

1 The fire resistance of a physical barrier of fire-resisting construction is a measure of its ability to withstand the effects of fire in one or more of the following ways:

(a) **integrity**: resistance to fire penetration, ie the prevention of the passage of flame and smoke;
(b) **insulation**: resistance to the transfer of excessive heat; and
(c) **load-bearing capacity**: resistance to collapse, ie to maintain support of the design load of other parts of the building, plant or structure, where the barrier also provides this.

2 The level of fire resistance is specified as the duration that the barrier is able to withstand the effects of fire in respect of one or more of these properties. This is determined through standard test procedures, where the periods of fire resistance are conventionally reported as:

(a) short (or 30 minutes);
(b) medium (or 60 minutes); or
(c) long (or 120 minutes).

3 The standard test procedures are those described in:

(a) BS 476 Parts 20–24;[66]
(b) BS EN 1363;[67] in conjunction with
(c) BS EN 1364 Parts 1–2;[68]
(d) BS EN 1365 Parts 1–4;[69]
(e) BS EN 1366 Parts 1–2;[70]
(f) BS EN 1634 Parts 1–2.[71]

Appendix 5 Fire reaction

1 The reaction to fire of a physical barrier of fire-resisting construction is a measure of the contribution the materials used in its construction have on the development and spread of the fire. This contribution is categorised as minimal, low, medium, high, and very high risk. The categories are determined in accordance with standard test procedures, though some materials are deemed to be of minimal risk and do not require to be tested. These are:

(a) concrete;
(b) fired clay (ie bricks);
(c) ceramics;
(d) steel;
(e) plaster and masonry containing not more than 1% by weight or volume of organic material; and
(f) concrete bricks or blocks.

2 The categories of materials that may be used in the construction of a physical barrier of fire-resisting construction are limited to: minimal, low, or medium. The minimum category that may be used depends on the type of barrier – this is discussed in paragraphs 266–279.

3 The test procedures and performance required by either the British Standards or the comparable European Harmonised Standards for the three categories are as follows:

	British Standards	**European Harmonised Standards**
Minimal	The material is certified non-combustible according to the test specified in BS 476-4[61] throughout, or the material does not flame or cause any rise in temperature on either the centre (specimen) or furnace thermocouples according to the test specified in BS 476-11[61]	The material has achieved a classification of A1 when tested in accordance with BS EN ISO 1182[72] and BS EN ISO 1716,[73] or the material has achieved a classification of A2 when tested in accordance with BS EN 13823[74] and BS EN ISO 1182[72] or BS EN ISO 1716[73]
Low	The surface of the material (or where it is bonded throughout to a substrate, the surface material combined with the substrate) has a surface spread of flame of Class 1 when tested in accordance with BS 476-7[61] and when tested in accordance with BS 476-6[61] has an index of performance (I) not exceeding 12 and a sub-index (i1) not exceeding 6	The material has achieved a classification of B or better when tested in accordance with BS EN 13823[74] and BS EN ISO 11925-1[75]
Medium	The material has a surface spread of flame of Class 1 when tested in accordance with BS 476-7[61]	The material has achieved a classification of C or better when tested in accordance with BS EN 13823[74] and BS EN ISO 11925-1[75]

Appendix 6 Workplace and work equipment transitional provisions under regulation 17

Arrangements for workplaces in use on or before 30 June 2003		
	When	**Action**
Equipment and protective systems already in use/available at the workplace	Immediately	Review equipment/protective systems against risk assessment requirements in regulation 5 of DSEAR. Equipment/protective systems at the workplace can continue to be used provided that the assessment indicates it is safe to do so.
Equipment and protective systems available for first time use after 30 June 2003	After 30 June 2003	Select equipment/protective systems that are new or available for the first time after this date in accordance with regulation 7(2)/Schedule 3 of DSEAR and the essential health and safety requirements of EPS. The place in which the equipment is to be located or used must be classified and zoned in accordance with regulation 7(1) of DSEAR if this has not yet been done during the transitional period for workplaces provided for in regulation 17(2)(a) of DSEAR.
Classification and zoning of hazardous areas	By 30 June 2006	Classify places into hazardous and non-hazardous places and zone hazardous places in accordance with regulation 7(1) of DSEAR. However, places may need to be classified before this date if equipment available for the first time after 30 June 2003 is to be used there.
Modifications etc to workplaces	After 30 June 2003	Ensure that any modification etc made to areas of the workplace that may contain an explosive atmosphere meets the requirements of regulations 7 and 11 of DSEAR from the date of the modification.
Marking hazardous places	By 30 June 2006	Provide any signs required by regulation 7(3)/Schedule 4 of DSEAR. If the part of the workplace to be marked is modified etc after 30 June 2003, but before 30 June 2006, regulation 17(3) of DSEAR requires that signs are provided from the date of the modification.
Provision of work clothing	By 30 June 2006	Provide antistatic work clothing as required by regulation 7(5) of DSEAR. However, if the part of the workplace in which it is to be worn is modified etc before this date it should be provided from the date of the modification.
Co-ordination of explosion protection measures	By 30 June 2006	Co-ordinate any measures required by regulation 11 of DSEAR and, as required by regulation 5(4) of DSEAR, record the aim of the co-ordination in the risk assessment. However, if part of the workplace is modified etc before 30 June 2006, the co-ordination requirements in respect of that part apply from the date of the modification.

Arrangements for workplaces used for the first time after 30 June 2003		
	When	**Action**
Equipment and protective systems	Immediately	Select equipment or protective systems etc in accordance with regulation 7(2)/Schedule 3 of DSEAR and Schedule 4 of EPS.
Classification and zoning of hazardous areas	Immediately	Classify places into hazardous and non-hazardous places and zone hazardous places in accordance with regulation 7(1) of DSEAR.
Modifications etc to workplaces	When modification is made	Ensure that any modifications etc made to areas of the workplace that may contain an explosive atmosphere meet the requirements of regulations 7(1), 7(3), 7(5) and 11 of DSEAR from the date of the modification.
Marking hazardous places	Immediately	Provide any signs required by regulation 7(3)/Schedule 4 of DSEAR.
Verification of explosion safety	Before workplace is used for the first time	Ensure that the overall explosion safety of the workplace is verified by someone who is competent to do so as required by regulation 7(4) of DSEAR.
Provision of work clothing	Immediately	Provide antistatic work clothing as required by regulation 7(5) of DSEAR.
Co-ordination of explosion protection measures	Immediately	Co-ordinate any measures required by regulation 11 of DSEAR.

Reference and further reading

References

1 *Controlling fire and explosion risks in the workplace: A brief guide to the Dangerous Substances and Explosive Atmospheres Regulations* Leaflet INDG370(rev1) HSE Books 2013 www.hse.gov.uk/pubns/indg370.htm

2 *Unloading petrol from road tankers. Dangerous Substances and Explosive Atmospheres Regulations 2002. Approved Code of Practice and guidance* L133 HSE Books 2003 ISBN 978 0 7176 2197 2 www.hse.gov.uk/pubns/books/l133.htm

3 BS EN 14034 Parts 1–4 British Standards Institution

- BS EN 14034-1:2004+A1:2011 *Determination of explosion characteristics of dust clouds. Determination of the maximum explosion pressure* p_{max} *of dust clouds*
- BS EN 14034-2:2006+A1:2011 *Determination of explosion characteristics of dust clouds. Determination of the maximum rate of explosion pressure rise* $(dp/dt)_{max}$ *of dust clouds*
- BS EN 14034-3:2006+A1:2011 *Determination of explosion characteristics of dust clouds. Determination of the lower explosion limit LEL of dust clouds*
- BS EN 14034-4:2004+A1:2011 *Determination of explosion characteristics of dust clouds. Determination of the limiting oxygen concentration LOC of dust clouds*

4 BS EN 13821:2002 *Potentially explosive atmospheres. Explosion prevention and protection. Determination of minimum ignition energy of dust/air mixtures* British Standards Institution

5 *Storage and handling of organic peroxides* Chemical Safety Guidance Note CS21 HSE Books 1991 ISBN 978 0 7176 2403 4 www.hse.gov.uk/pubns/books/cs21.htm

6 BS EN 61508-1:2010 *Functional safety of electrical/electronic/programmable electronic safety-related systems. General requirements* British Standards Institution

7 BS EN 61511-2:2004 *Functional safety. Safety instrumented systems for the process industry sector. Guidelines for the application of IEC 61511-1* British Standards Institution

8 *Fire safety in construction: Guidance for clients, designers and those managing and carrying out construction work involving significant fire risks* HSG168 (Second edition) HSE Books 2010 ISBN 978 0 7176 6345 3 www.hse.gov.uk/pubns/books/hsg168.htm

9 Code of practice 7 *Storage of Full and Empty LPG Cylinders and Cartridges* UKLPG March 2004 www.uklpg.org

10 *Guidance for the storage of gas cylinders in the workplace* British Compressed Gas Association GN 2 www.bcga.co.uk

11 *Guidance for the storage of liquids in intermediate bulk containers* A joint publication of the Chemical Business Association (CBA) and Solvents Industry Association (SIA) March 2008 www.chemical.org.uk/regulatoryissues/healthandsafety/jointcbasiastorageguidance.aspx

12 Model Code of Safe Practice Part 15 *Area Classification Code for Installations Handling Flammable Fluids* (Third edition) Energy Institute July 2005 www.energypublishing.org/

13 *The storage of flammable liquids in containers* HSG51 (Second edition) HSE Books 1998 ISBN 978 0 7176 1471 4 www.hse.gov.uk/pubns/books/hsg51.htm

14 *The storage of flammable liquids in tanks* HSG176 HSE Books 1998 ISBN 978 0 7176 1470 7 www.hse.gov.uk/pubns/books/hsg176.htm

15 *The safe use and handling of flammable liquids* HSG140 HSE Books 1996 ISBN 978 0 7176 0967 3 www.hse.gov.uk/pubns/books/hsg140.htm

16 BS EN 14470-1:2004 *Fire safety storage cabinets. Safety storage cabinets for flammable liquids* British Standards Institution

17 NFPA 30: *Flammable and Combustible Liquids Code* www.nfpa.org/catalog/product.asp?pid=NFPA30ARCHIV&cookie%5Ftest=1

18 BS EN 14470-2:2006 *Fire safety storage cabinets. Safety cabinets for pressurised gas cylinders* British Standards Institution

19 *Safety of pressure systems: Pressure Systems Safety Regulations 2000. Approved Code of Practice* L122 HSE Books 2000 ISBN 978 0 7176 1767 8 www.hse.gov.uk/pubns/books/l122.htm

20 *Guidance for the location and design of occupied buildings on chemical manufacturing sites* (Third edition) Chemical Industries Association www.cia.org.uk

21 BS EN ISO 16852:2010 *Flame arresters. Performance requirements, test methods and limits for use* British Standards Institution

22 BS 5925:1991 *Code of practice for ventilation principles and designing for natural ventilation* British Standards Institution

23 *Natural ventilation in non-domestic buildings* AM10 CIBSE Applications Manual 2005

24 *Environmental design* CIBSE Guide A 2006

25 *Clearing the air: A simple guide to buying and using local exhaust ventilation (LEV)* Leaflet INDG408 HSE Books 2008 www.hse.gov.uk/pubns/indg408.htm

26 *Time to clear the air! A workers' pocket guide to local exhaust ventilation (LEV)* INDG409 HSE Books 2008 ISBN 978 0 7176 6300 2 www.hse.gov.uk/pubns/indg409.htm

27 *Controlling airborne contaminants at work: A guide to local exhaust ventilation (LEV)* HSG258 (Second edition) HSE Books 2011 ISBN 978 0 7176 6415 3 www.hse.gov.uk/pubns/books/hsg258.htm

28 Code of Practice 1 Part 1 *Bulk LPG Storage at Fixed Installations: Design, Installation and Operation of Vessels Located Above Ground* UKLPG www.uklpg.org

29 Code of Practice 12 *Recommendations for Safe Practice in the Design and Operation of LPG Cylinder Filling Plants* UKLPG www.uklpg.org

30 *Safety and environmental standards for fuel storage sites* Process Safety Leadership Group 2009 ISBN 978 0 7176 6386 6 www.hse.gov.uk/comah/buncefield/fuel-storage-sites.pdf

31 *Secondary and Tertiary Containment of Bulk Hazardous Liquids at COMAH Establishments* www.hse.gov.uk/comah/guidance/bulk-hazardous-liquids.pdf HSE/Environment Agency paper 2010

32 BS EN 13463-6:2005 *Non-electrical equipment for potentially explosive atmospheres. Protection by control of ignition source 'b'* British Standards Institution

33 BS 5958-1:1991 *Code of Practice for control of undesirable static electricity. General considerations* British Standards Institution

34 PD CLC/TR 50404:2003 *Electrostatics. Code of Practice for the avoidance of hazards due to static electricity* British Standards Institution

35 BS EN 14986:2007 *Design of fans working in potentially explosive atmospheres* British Standards Institution

36 *Chemical warehousing: The storage of packaged dangerous substances* HSG71 (Fourth edition) HSE Books 2009 ISBN 978 0 7176 6237 1 www.hse.gov.uk/pubns/books/hsg71.htm

37 *Safe handling of combustible dusts: Precautions against explosions* HSG103 (Second edition) HSE Books 2003 ISBN 978 0 7176 2726 4 www.hse.gov.uk/pubns/books/hsg103.htm

38 BS EN 14797:2006 *Explosion venting devices* British Standards Institution

39 BS EN 14491:2012 *Dust explosion venting protective systems* British Standards Institution

40 BS EN 14994:2007 *Gas explosion venting protective systems* British Standards Institution

41 BS EN 14373:2005 *Explosion suppression systems* British Standards Institution

42 BS EN 14460:2006 *Explosion resistant equipment* British Standards Institution

43 BS EN 15089:2009 *Explosion isolation systems* British Standards Institution

44 BS EN 16009:2011 *Flameless explosion venting devices* British Standards Institution

45 BS EN 16020:2011 *Explosion diverters* British Standards Institution

46 BS EN ISO 16852:2010 *Flame arresters: Performance requirements, test methods and limits for use* British Standards Institution

47 *Guidelines on the application Directive 94/9/EC of the European Parliament and Council of 23 March 1994 on the approximation of the laws of the member states concerning Equipment and Protective Systems intended for use in potentially Explosive Atmospheres* Third edition June 2009, updated May 2011 http://ec.europa.eu/enterprise/sectors/mechanical/files/atex/guide/atexguidelines-may2011_en.pdf

48 BS 5908-2:2012 *Fire and explosion precautions at premises handling flammable gases, liquids and dusts. Guide to applicable standards and regulations* British Standards Institution

49 *Designing and operating safe chemical reaction processes* HSG143 HSE Books 2000 ISBN 978 0 7176 1051 8 www.hse.gov.uk/pubns/books/hsg143.htm

50 Model Code of Safe Practice Part 16 *Tank cleaning safety code* (Third edition) July 2008 ISBN 978 0 85293 436 4 Energy Institute

51 *Guidance on permit-to-work systems: A guide for the petroleum, chemical and allied industries* HSG250 HSE Books 2005 ISBN 978 0 7176 2943 5 www.hse.gov.uk/pubns/books/hsg250.htm

52 *Safe work in confined spaces. Confined Spaces Regulations 1997. Approved Code of Practice, Regulations and guidance* L101 (Second edition) HSE Books 2009 ISBN 978 0 7176 6233 3 www.hse.gov.uk/pubns/books/l101.htm

53 *The safe use of oxy-fuel gas equipment (individual portable or mobile cylinder supply)* BCGA CP 7 British Compressed Gas Association www.bcga.co.uk

54 *Cryogenic liquid storage at users' premises* BCGA CP 36 British Compressed Gas Association www.bcga.co.uk

55 *DSEAR Risk Assessment* BCGA GN 13 British Compressed Gas Association www.bcga.co.uk

56 *The safe handling of gas cylinders at waste facilities* BCGA L 2 British Compressed Gas Association www.bcga.co.uk

57 Code of Practice 26 *Uplifting of Static LPG Vessels from Site and their Carriage to and from Site by Road* (November 2004) incorporating Amendment 1 2007 UKLPG www.uklpg.org

58 *Guidance for Design, Construction, Modification, Maintenance and Decommissioning of Filling Stations* (The Blue Book) APEA www.apea.org.uk-guidance www.apea.org.uk/publication/blue-book-pdf

59 BS EN 60079-10-1:2009 *Explosive atmospheres. Classification of areas. Explosive gas atmospheres* British Standards Institution

60 BS EN 60079-10-2:2009 *Explosive atmospheres. Classification of areas. Combustible dust atmospheres* British Standards Institution

61 BS 476 Parts 4, 6, 7, 11 British Standards Institution

- BS 476-4:1970 *Fire tests on building materials and structures. Non-combustibility test for materials*
- BS 476-6:1989+A1:2009 *Fire tests on building materials and structures. Method of test for fire propagation for products*
- BS 476-7:1997 *Fire tests on building materials and structures. Method of test to determine the classification of the surface spread of flame of products*
- BS 476-11:1982 *Fire tests on building materials and structures. Method for assessing the heat emission from building materials*

62 BS EN 13463-1:2009 *Non-electrical equipment for use in potentially explosive atmospheres. Basic method and requirements* British Standards Institution

63 BS EN 60079-14:2008 *Explosive atmospheres. Electrical installations design, selection and erection* British Standards Institution

64 BS EN 60079-0:2012 *Explosive atmospheres. Equipment. General requirements* British Standards Institution

65 BS 5958-2:1991 *Code of practice for control of undesirable static electricity. Recommendations for particular industrial situations* British Standards Institution

66 BS 476 Parts 20–24 *Fire test on building materials and structures* British Standards Institution

- BS 476-20:1987 *Method for determination of the fire resistance elements of construction (general principles)*
- BS 476-21:1987 *Methods for determination of the fire resistance of loadbearing elements of construction*
- BS 476-22:1987 *Methods for determination of the fire resistance of non-loadbearing elements of construction*
- BS 476-23:1987 *Methods for determination of the contribution of components to the fire resistance of a structure*
- BS 476-24:1987 *Method for determination of the fire resistance of ventilation ducts*

67 BS EN 1363-1:2012 *Fire resistance tests. General requirements* British Standards Institution

68 BS EN 1364 Parts 1–2 British Standards Institution

- BS EN 1364-1:1999 *Fire resistance tests for non-loadbearing elements. Walls*
- BS EN 1364-2:1999 *Fire resistance tests for non-loadbearing elements. Ceilings*

69 BS EN 1365 Parts 1–4 British Standards Institution

- BS EN 1365-1:2012 *Fire resistance tests for loadbearing elements. Walls*
- BS EN 1365-2:2000 *Fire resistance tests for loadbearing elements. Floors and roofs*
- BS EN 1365-3:2000 *Fire resistance tests for loadbearing elements. Beams*
- BS EN 1365-4:1999 *Fire resistance tests for loadbearing elements. Columns*

70 BS EN 1366 Parts 1–2 British Standards Institution

- BS EN 1366-1:1999 *Fire resistance tests for service installations. Ducts*
- BS EN 1366-2:1999 *Fire resistance tests for service installations. Fire dampers*

71 BS EN 1634 Parts 1–2 British Standards Institution

- BS EN 1634-1:2008 *Fire resistance and smoke control tests for door, shutter and, openable window assemblies and elements of building hardware. Fire resistance tests for doors, shutters and openable windows*
- BS EN 1634-2:2008 *Fire resistance and smoke control tests for door, shutter and openable window assemblies and elements of building hardware. Fire resistance characterisation test for elements of building hardware*

72 BS EN ISO 1182:2010 *Reaction to fire tests for products. Non-combustibility test* British Standards Institution

73 BS EN ISO 1716:2010 *Reaction to fire tests for products. Determination of the gross heat of combustion (calorific value)* British Standards Institution

74 BS EN 13823:2010 *Reaction to fire tests for building products. Building products excluding floorings exposed to the thermal attack by a single burning item* British Standards Institution

75 BS ISO TR 11925-1:1999 *Reaction to fire tests. Ignitability of building products subjected to direct impingement of flame. Guidance on ignitability* British Standards Institution

Further reading

The EU has produced a general advisory guide but note that its requirement to produce an Explosion Protection Document is not applicable in GB because the information is already covered domestically by DSEAR Regulations 5, 6 and 7:

- http://bookshop.europa.eu/is-bin/INTERSHOP.enfinity/WFS/EU-Bookshop-Site/en_GB/-/EUR/ViewPublication-Start?PublicationKey=KE6404175

Glossary

ACOP Approved Code of Practice.

ADR European Agreement concerning the International Carriage of Dangerous Goods by Road.

ALARP as low as reasonably practicable.

ambient temperature the temperature of an immediate locality.

ATEX the collective name for the Explosive Atmospheres Directives 99/92/EC and 94/9/EC of the European Parliament. Both deal with precautions for explosive atmospheres. The first (also known as ATEX 137) focuses on worker safety and the second ('ATEX 95' or 'the ATEX Equipment Directive') focuses on the correct standard of equipment to avoid ignition. Also see EPS.

auto-ignition temperature the minimum temperature at which a material will ignite spontaneously under specified test conditions. Also referred to as the minimum ignition temperature.

CAD Chemical Agents Directive 98/24/EC.

CDG Carriage of Dangerous Goods and Use of Transportable Pressure Equipment Regulations 2009.

CHIP Chemicals (Hazard Information and Packaging for Supply) Regulations 2009.

CLP European Classification, Labelling and Packaging of Substances and Mixtures Regulation 2008.

COMAH Control of Major Accident Hazards Regulations 1999.

combustible capable of burning in air when ignited.

control steps taken or provisions put in place to reduce the likelihood of a fire, explosion, or similar event happening.

COSHH Control of Substances Hazardous to Health Regulations 2002.

DSEAR the Dangerous Substances and Explosive Atmospheres Regulations 2002.

EEA European Economic Area.

EHSRs essential heath and safety requirements (as set out in EU requirements such as the EPS Regulations).

enforcing authority the authority with responsibility for enforcing the HSW Act and other relevant statutory provisions.

EPS Equipment and Protective Systems for use in Potentially Explosive Atmospheres Regulations 1996 implementing Directive 94/9/EC of the European Parliament.

flame arrester a device consisting of an element, a housing and associated fittings which is constructed and used to prevent the passage of flame.

flammable capable of burning with a flame

flammable liquid for the purpose of this book, flammable liquid means a liquid with a flashpoint of 60 °C or below and stored at a near atmospheric pressure.

flashpoint the minimum temperature at which a liquid, under specific test conditions, gives off sufficient flammable vapour to ignite momentarily on the application of an ignition source.

fire resistance the ability of a material, product, assembly or structure to fulfil, for a stated period of time, the required stability against the passage of flame and hot gases, and if additionally specified, thermal insulation and/or load-bearing capacity in a standard fire resistance test.

fire-resisting ability to fulfil, for a stated period of time, the required stability, fire integrity and/or thermal insulation, where appropriate, in a standard fire resistance test.

fire wall an imperforate wall, screen or partition capable of affording at least 30 minutes fire resistance, if tested in accordance with the appropriate standard against the passage of flame or heat.

hazard something that could cause harm to people.

hazardous area an area where flammable or explosive gas, or vapour–air mixtures (often referred to as explosive gas–air mixtures) are, or may be expected to be, present in quantities which require special precautions to be taken against the risk of ignition.

hot work this includes welding or the use of any equipment likely to cause flame, sparks or heat.

HSW Act Health and Safety At Work etc Act 1974.

incendive having sufficient energy to ignite a flammable mixture.

inert incapable of supporting combustion; to render incapable of supporting combustion.

LEV local exhaust ventilation.

lower explosion limit (LEL) the minimum concentration of vapour in air below which propagation of a flame will not occur in the presence of an ignition source. Also referred to as the lower flammable limit or lower explosive limit.

LPG Liquefied Petroleum Gas.

Management Regulations Management of Health and Safety at Work Regulations 1999.

MEV mechanical extract or exhaust ventilation.

mitigation the steps taken to minimise the consequences during and after the occurrence of a fire, explosion or similar event.

NRW Natural Resources Wales.

ONR Office for Nuclear Regulation.

permit-to-work a document issued by an authorised person to permit work to be carried out safely in a defined area under specified conditions.

PPE personal protective equipment (clothing, face protection etc).

PSSR Pressure Systems Safety Regulations 2000.

PUWER the Provision and Use of Work Equipment Regulations 1992.

REACH Registration, Evaluation, Authorisation and restriction of Chemicals Regulations EC 1907/2006.

RID Regulations concerning the International Carriage of Dangerous Goods by Rail.

risk the chance – however large or small – that a hazard could cause harm.

ROSOV remotely operated shut-off valve.

RPE respiratory protective equipment (filtering face masks).

SEPA Scottish Environment Protection Agency.

SDS safety data sheet (formerly known as material safety data sheets).

underground tank a tank buried in the ground so that no part of the tank is above ground except for fittings attached to the tank.

upper explosion limit the maximum concentration of vapour in air above which the propagation of a flame will not occur. Also referred to as the upper flammable limit or the upper explosion limit.

vapour the gaseous phase released by evaporation from a material that is a liquid at normal temperatures and pressure.

zone the classified part of a hazardous area, representing the probability of a flammable vapour (or gas) and air mixture being present.

Further information

For information about health and safety visit https://books.hse.gov.uk or http://www.hse.gov.uk. You can view HSE guidance online and order priced publications from the website. HSE priced publications are also available from bookshops.

To report inconsistencies or inaccuracies in this guidance email: commissioning@wlt.com.

British Standards can be obtained in PDF or hard copy formats from BSI: http://shop.bsigroup.com or by contacting BSI Customer Services for hard copies only Tel: 0846 086 9001 email: cservices@bsigroup.com.

The Stationery Office publications are available from The Stationery Office, PO Box 29, Norwich NR3 1GN Tel: 0333 202 5070 Fax: 0333 202 5080. E-mail:customer.services@tso.co.uk Website: www.tso.co.uk. They are also available from bookshops.

Statutory Instruments can be viewed free of charge at www.legislation.gov.uk where you can also search for changes to legislation.